OPERATING SYSTEMS

Dr. K. DEEBA
Ms. L. D. SUJITHRA DEVI

SHANLAX PUBLICATIONS
66,61, Vasantha Nagar,
Madurai – 625003
e-mail: shanlaxpublications@gmail.com
web: www.shanlaxpublications.com

Operating Systems

ISBN: 978-93-85977-30-5

Price: ₹ 490 /-

Publisher

SHANLAX PUBLICATIONS

61, 66 T.P.K. Main Road,
Vasantha Nagar, Madurai - 625003
Ph: 0452-4208765, 7200303383
email: shanlaxpublications@gmail.com
web: www.shanlaxpublications.com

PREFACE

The purpose of the book is to explore the knowledge of the reader to the basic concepts of **Operating Systems** in line with the syllabi prescribed by the Anna University-Chennai. This book is designed to help the students to understand the subject easily and prepare for the University Examinations. The chapters in the book are clearly understandable for the students in such a way that the concepts are easily mentioned.

Review questions are given at the end of each chapter. Review questions are separated as short answer questions and essay type questions. Each chapter is explained with illustrative example problems and diagrammatically represented wherever necessary.

We wish to convey our gratitude to the entire team of publications for taking all pains in the publication of the book in time with quality printing.

Useful suggestions and healthy criticism to improve this book will be welcomed and thankfully acknowledged.

Dr. K. DEEBA
Ms. L. D. SUJITHRA DEVI

ACKNOWLEDGEMENT

Foremost we thank the Lord Almighty, for giving us the courage, strength and knowledge to undertake this task.

Our sincere thanks to the **Management of Hindusthan College of Engineering and Technology**, **Coimbatore** for providing necessary facilities to carry out the work.

It is our immense pleasure to express our sincere and heartfelt thanks to **Dr. T. Kannadasan**, Principal, Hindusthan College of Engineering and Technology, for his constant encouragement, which had a thoughtful effect in completing the work.

It is our gratitude to thank our Director- Academics and Head of the Department, **Dr. N. Sengottaiyan,** Computer Science and Engineering for his invaluable moral support at every stage of our work.

We thank all the faculty members of Computer Science and Engineering Department, Hindusthan College of Engineering and Technology, for their kind cooperation and useful suggestions.

We also thank our family members for their support and encouragement.

Dr. K. DEEBA
Ms. L. D. SUJITHRA DEVI

UNIT I
OPERATING SYSTEMS OVERVIEW

UNIT II
PROCESS MANAGEMENT

UNIT III
STORAGE MANAGEMENT

UNIT IV

I/O SYSTEMS

UNIT-5
CASE STUDY

1

COMPUTER SYSTEM OVERVIEW

1.1 OPERATING SYSTEM OVERVIEW

A program that acts as an intermediary between a user of a computer and the computer hardware is operating system.

Operating system goals:

Execute user programs and make solving user problems easier.

Make the computer system convenient to use.

Use the computer hardware in an efficient manner.

Computer System Components:

Hardware – provides basic computing resources (CPU, memory, I/O devices).

Operating system – controls and coordinates the use of the hardware among the various application programs for the various users.

Applications programs – define the ways in which the system resources are used to solve the computing problems of the users (compilers, database systems, video games, business programs).

Users (people, machines, other computers).

The design of a new operating system is a major task. It is important that the goals of the system be well defined before the design begins. Because an operating system is large and complex, it must be created piece by piece. Each of these pieces should be a well-delineated portion of the system, with carefully defined inputs, outputs, and functions.

It exploits the hardware resources of one or more processors to provide a set of services to system users and manages secondary memory and I/O devices

1.1.1 Basic Elements

Four main structural elements:

- **Processor:**
  - Controls the operation of the computer
  - Performs its data processing functions.
  - It is often referred to as the central processing unit (CPU).

- **Main memory:**
  - Stores data and programs.
  - Volatile
  - Contents of the memory is lost when the computer is shut down
  - Referred to as real memory or primary memory.

- **I/O modules:**
  - Move data between the computer and its external environment.
  - The external environment consists of a variety of devices, including secondary memory devices (e.g., disks), communications equipment, and terminals.

- **System bus:**
  - Provides for communication among processors, main memory, and I/O modules.

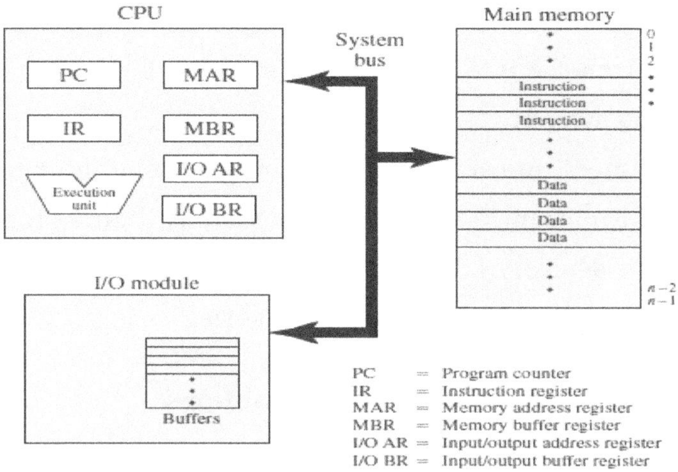

Figure 1.1 Computer Components: Top-Level View

TOP-LEVEL COMPONENTS

Processor Module:

One of the processor's functions is to exchange data with memory. For this purpose, it uses two internal (to the processor) registers:

Memory Address Register (MAR), which specifies the address in memory for the next read or write;

Memory Buffer Register (MBR), which contains the data to be written into memory or which receives the data read from memory.

I/O Module:

An I/O address register (I/OAR) specifies a particular I/O device.

An I/O buffer register (I/OBR) is used for the exchange of data between an I/O module and the processor.

Memory Module:

A memory module consists of a set of locations, defined by sequentially numbered addresses. Each location contains a bit pattern that can be interpreted as either an instruction or data. An I/O module transfers data from external devices to processor and memory, and vice versa. It contains internal buffers for temporarily holding data until they can be sent on.

EVOLUTION OF OS

Serial Processing (From the late 1940s to the mid-1950s)

- ➢ No operating system
  - ❖ programmers interacted directly with the computer hardware
- ➢ Computers ran from a console with display lights, toggle switches, some form of input device, and a printer
- ➢ Users have access to the computer in "series"

These early systems presented two main problems:

- **Scheduling**
  - ➢ most installations used a hardcopy sign-up sheet to reserve computer time
  - ➢ time allocations could run short or long, resulting in wasted computer time
- **Setup time**
  - ➢ A considerable amount of time was spent just on setting up the program to run.

EVOLUTION OF MICROPROCESSOR

1. **Microprocessor**
   - Invention that brought about desktop and handheld computing
   - Processor on a single chip
   - Fastest general purpose processor
   - Multiprocessors
   - Each chip (socket) contains multiple processors (cores) each with multiple levels of large memory caches, and multiple logical processors haring the execution units of each core.
2. **Graphical Processing Units (GPUs)**
   - Provide efficient computation on arrays of data using Single-Instruction Multiple Data (SIMD) techniques
   - Used for general numerical processing
   - Physics simulations for games
   - Computations on large spreadsheets

3. Digital Signal Processors (DSPs)

- Deal with streaming signals such as audio or video
- Used to be embedded in devices like modems
- Encoding/decoding speech and video (codecs)
- Support for encryption and security

4. System on a Chip (SoC)

- To satisfy the requirements of handheld devices, the microprocessor is giving way to the SoC
- Components such as DSPs, GPUs, codecs and main memory, in addition to the CPUs and caches, are on the same chip

1.2 INSTRUCTION EXECUTION

Program to be executed by a processor consists of a set of instructions stored in memory. It consists of two steps:

- ➢ The processor reads (fetches) instructions from memory one at a time and executes each instruction.
- ➢ Program execution consists of repeating the process of instruction fetch and instruction execution.

Program execution consists of repeating the process of instruction fetch and instruction execution.

1. Instruction Fetch and Execute

- The processing required for a single instruction is called an instruction cycle. Using a simplified two-step description, the instruction cycle is depicted in Figure 1.2.

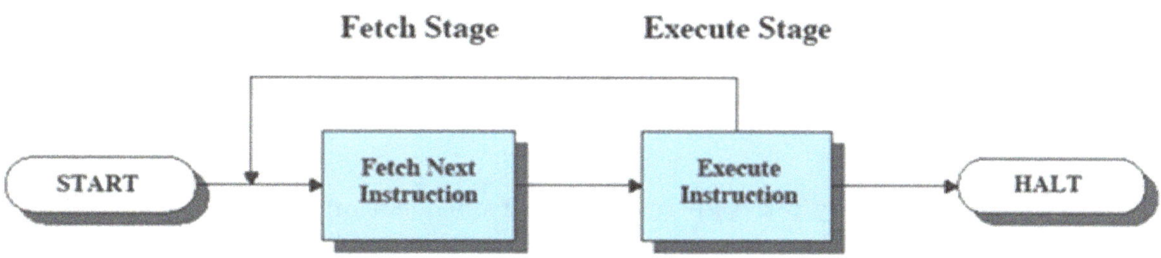

Figure 1.2 Basic Instruction Cycle

- The two steps are referred to as the fetch stage and the execute stage. Program execution halts only if the processor is turned off, some sort of unrecoverable error occurs, or a program instruction that halts the processor is encountered.
- The processor fetches the instruction from memory
- Program counter (PC) holds address of the instruction to be fetched next
  - ➢ PC is incremented after each fetch (i.e., the instruction located at the next higher memory address).

For example

Consider a simplified computer in which each instruction occupies one16-bit word of memory. Assume that the program counter is set to location 300.The processor will next fetch the instruction at location 300. On succeeding instruction cycles, it will fetch instructions from locations 301, 302, 303, and so on. This sequence may be altered.

The fetched instruction is loaded into the Instruction Register (IR)

- ➢ The instruction contains bits that specify the action the processor is to take.
- ➢ The processor interprets the instruction and performs the required action,

which fall into four categories:

- **Processor-memory**: Data may be transferred from processor to memory or from memory to processor.
- **Processor-I/O**: Data may be transferred to or from a peripheral device by transferring between the processor and an I/O module.
- **Data processing**: The processor may perform some arithmetic or logic operation on data.
- **Control:** An instruction may specify that the sequence of execution be altered.

Characteristics of a Hypothetical Machine:

For example

The processor may fetch an instruction from location 149, which specifies that the next instruction will be from location 182. The processor sets the program counter to 182. Thus, on the next fetch stage, the instruction will be fetched from location 182 rather than 150.

An instruction's execution may involve a combination of these actions.

Consider a simple example using a hypothetical processor that includes the characteristics listed in Figure 1.3.

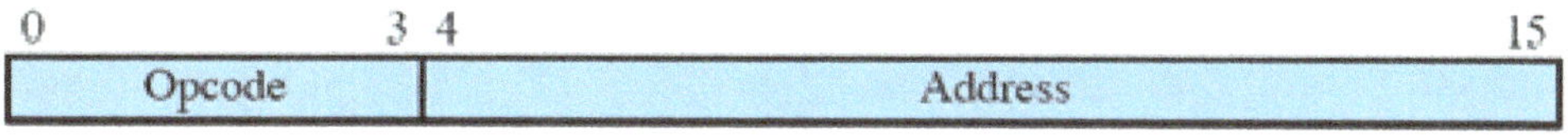

(a) Instruction format

(b) Integer format

Program counter (PC) = Address of instruction
Instruction register (IR) = Instruction being executed
Accumulator (AC) = Temporary storage

(c) Internal CPU registers

0001 = Load AC from memory
0010 = Store AC to memory
0101 = Add to AC from memory

(d) Partial list of opcodes

Figure 1.3 Characteristics of a Hypothetical Machine

- The processor contains a single data register, called the accumulator (AC). Both instructions and data are 16 bits long, and memory is organized as a sequence of 16-bit words.
- The instruction format provides 4 bits for the opcode, allowing as many as 2^4= 16 different opcodes (represented by a single hexadecimal 1 digit). The opcode defines the operation the processor is to perform. With the remaining 12 bits of the instruction format, up to2^{12} =4,096 (4K) words of memory (denoted by three hexadecimal digits) can be directly addressed.

Example of Program Execution:

Figure 1.4 illustrates a partial program execution, showing the relevant portions of memory and processor registers. The program fragment shown adds the contents of the memory word at address 940 to the contents of the memory word at address 941 and stores the result in the latter location. Three instructions, which can be described as three fetch and three execute stages, are required:

1. The PC contains 300, the address of the first instruction. This instruction (the value 1940 in hexadecimal) is loaded into the IR and the PC is incremented.
   - This process involves the use of a memory address register (MAR) and a memory buffer register (MBR). For simplicity, these intermediate registers are not shown.

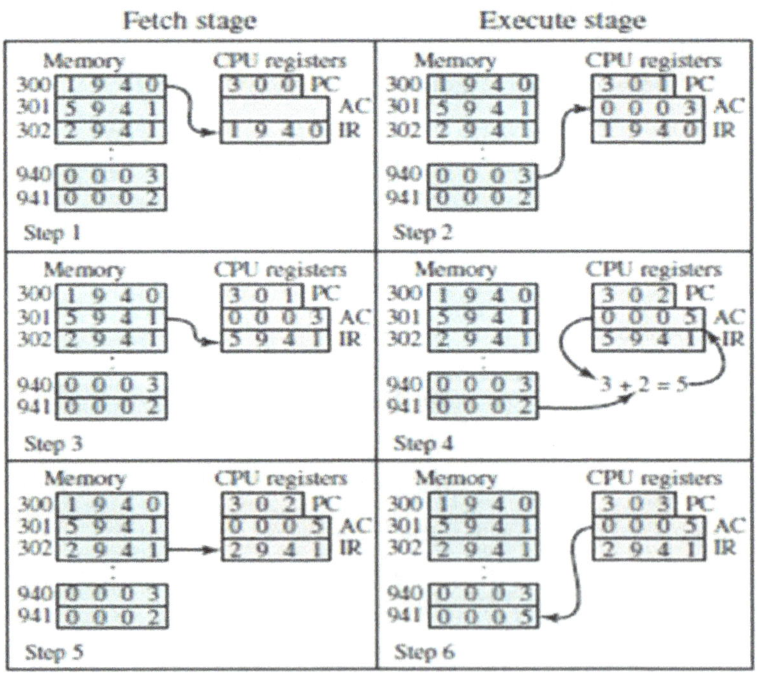

Figure 1.4 Example of Program Execution (contents of memory and registers in hexadecimal)

2. The first 4 bits (first hexadecimal digit) in the IR indicate that the AC is to be loaded from memory. The remaining 12 bits (three hexadecimal digits) specify the address, which is 940.
3. The next instruction (5941) is fetched from location 301 and the PC is incremented.
4. The old contents of the AC and the contents of location 941 are added and the result is stored in the AC.
5. The next instruction (2941) is fetched from location 302 and the PC is incremented.
6. The contents of the AC are stored in location 941.

In this example, three instruction cycles, each consisting of a fetch stage and an execute stage, are needed to add the contents of location 940 to the contents of 941.

With a more complex set of instructions, fewer instruction cycles would be needed. Most modern processors include instructions that contain more than one address. Thus the execution stage for a particular instruction may involve more than one reference to memory.

Also, instead of memory references, an instruction may specify an I/O operation.

1.3 INTERRUPTS

Computers provide a mechanism by which other modules (I/O, memory) may interrupt the normal sequencing of the processor. Interrupts are provided primarily to improve processor utilization.

Classes of Interrupts:

- **Program** - Generated by some condition that occurs as a result of an instruction execution, such as arithmetic overflow, division by zero, attempt to execute an illegal machine instruction, and reference outside a user's allowed memory space.
- **Timer** - Generated by a timer within the processor. This allows the operating system to perform certain functions on a regular basis.
- **I/O** - Generated by an I/O controller, to signal normal completion of an operation or to signal a variety of error conditions.
- **Hardware failure** - Generated by a failure, such as power failure or memory parity error.

The I/O program consists of three sections:

- **A sequence of instructions**, labeled 4 in the figure, to prepare for the actual I/O operation. i.e., copying the data to be output into a special buffer and preparing the parameters for a device command.
- **The actual I/O command**. Without the use of interrupts, once this command is issued, the program must wait for the I/O device to perform the requested function.
- **A sequence of instructions**, labeled 5 in the figure, to complete the operation, which includes setting a flag indicating the success or failure of the operation.

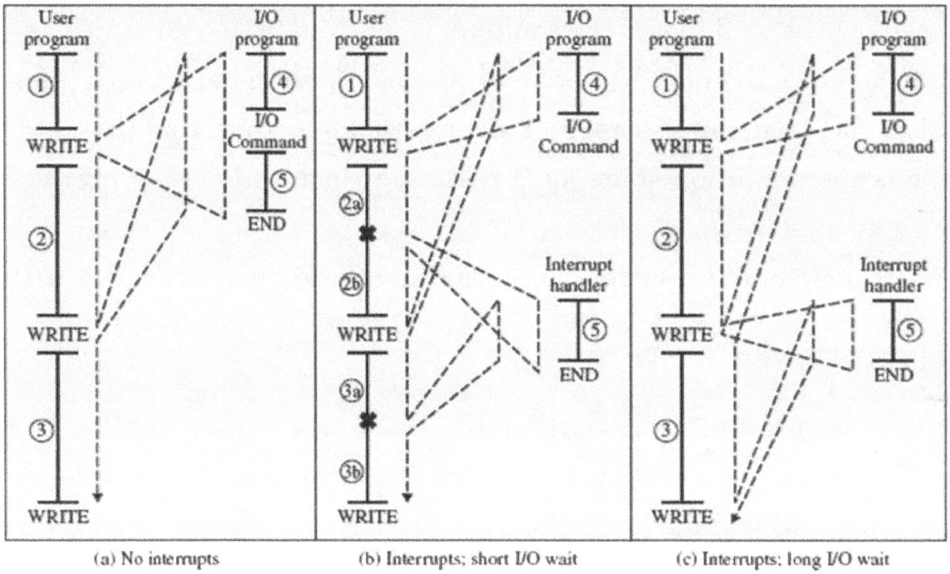

Figure 1.5 Program Flow of Control without and with Interrupts

Flow of Control:

a. **No Interrupts**

- The user program performs a series of WRITE calls interleaved with processing. The solid vertical lines represent segments of code in a program.
- Code segments 1, 2, and 3 refer to sequences of instructions that do not involve I/O.
- The WRITE calls are to an I/O routine that is a system utility and that will perform the actual I/O operation.

b. Interrupts: Short I/O Wait

- The dashed line represents the path of execution followed by the processor
- After the first WRITE instruction is encountered, the user program is interrupted and execution continues with the I/O program.

  After the I/O program execution is complete, execution resumes in the user immediately following the WRITE instruction.
- The I/O operation may take a relatively long time to complete, the I/O program is hung up waiting for the operation to complete; hence, the user program is stopped at the point of the WRITE call for some period of time.

Timing Diagram Based on the Flow of Control in Figures 1.5a and 1.5b .

Program Timing: Short I/O Wait

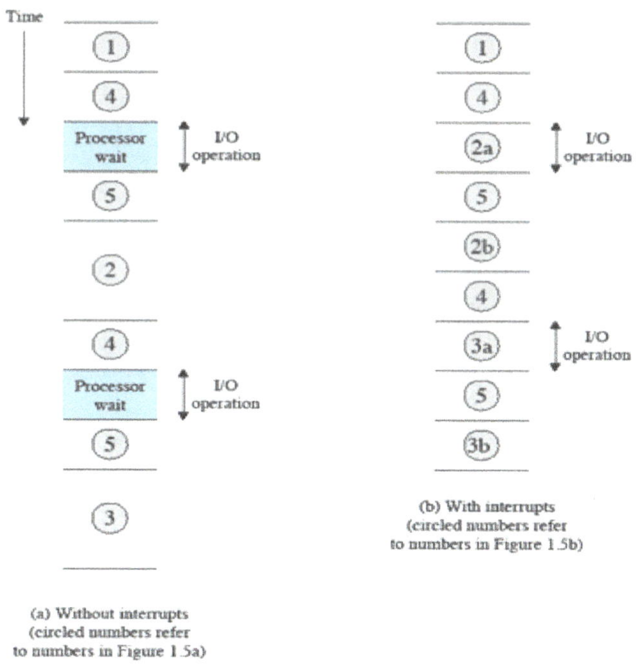

Figure 1.6 Program Timing: Short I/O Wait

Figures 1.5b and 1.8 assume that the time required for the I/O operation is relatively short: less than the time to complete the execution of instructions between write operations in the user program.

Example: printer

Interrupts: Long I/O Wait: In Figure 1.5c

The user program reaches the second WRITE call before the I/O operation spawned by the first call is complete. The result is that the user program is hung up at that point. When the preceding I/O operation is completed, this new WRITE call may be processed, and a new I/O operation may be started.

Figure 1.9 shows the timing for this situation with and without the use of interrupts.

Program Timing: Long I/O Wait

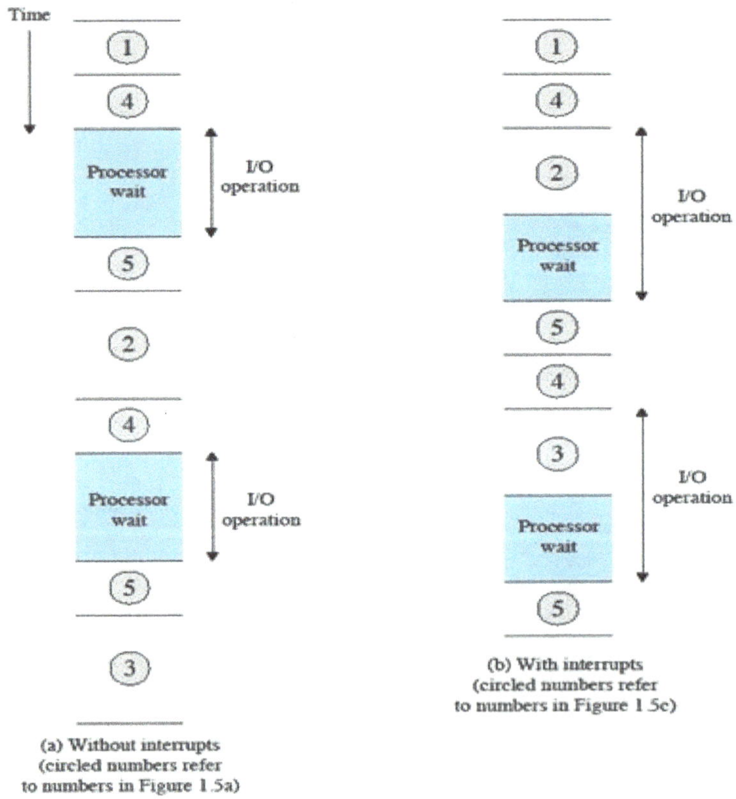

Figure 1.9 Program Timing: Long I/O Wait

1.3.1 Types of Interrupts

1. Single interrupts
2. Multiple interrupts

1. Single Interrupts

Interrupts and the Instruction Cycle:

- With interrupts, the processor executes other instructions while an I/O operation is in progress.

Consider the flow of control in Figure 1.5b:

The I/O program that is invoked in this case consists only of the preparation code and the actual I/O command.

- After these few instructions have been executed, control returns to the user program. Meanwhile, the external device is busy accepting data from computer memory and printing it. This I/O operation is conducted concurrently with the execution of instructions in the user program.
- When the external device becomes ready to be serviced, that is, when it is ready to accept more data from the processor, the I/O module for that external device sends an interrupt request signal to the processor. The processor responds by suspending operation of the current program; branching off to a routine to service.

1. Transfer of Control via Interrupts

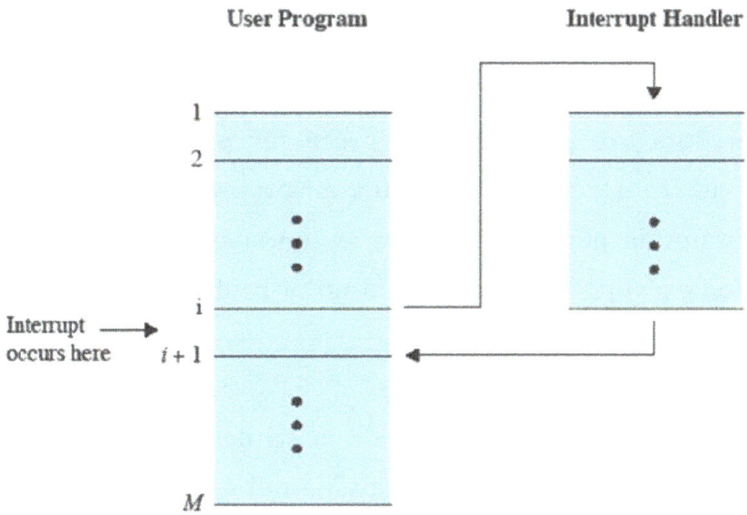

Figure 1.10 Transfer of Control via Interrupts

- For the user program, an interrupt suspends the normal sequence of execution. Hence the interrupt processing is completed, execution resumes (Figure 1.6).
- Thus, the user program does not have to contain any special code to accommodate interrupts; the processor and the OS are responsible for suspending the user program and then resuming it at the same point.

To accommodate interrupts, an interrupt stage is added to the instruction cycle, as shown in Figure 1.7.

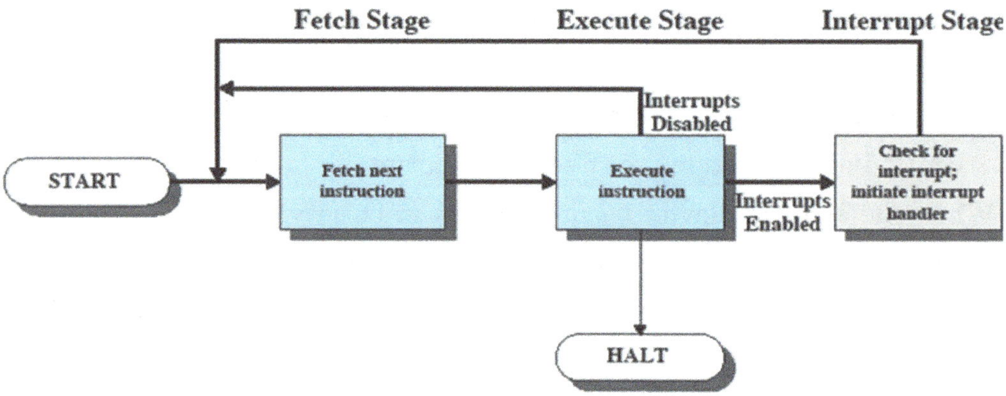

Figure 1.11 Instruction Cycle with Interrupts

In the interrupt stage, the processor checks to see if any interrupts have occurred, indicated by the presence of an interrupt signal.

- If no interrupts are pending, the processor proceeds to the fetch stage and fetches the next instruction of the current program.
- If an interrupt is pending, the processor suspends execution of the current program and executes an interrupt handler routine.

2. Multiple Interrupts

Suppose one or more interrupts can occur while an interrupt is being processed.

For example, a program may be receiving data from a communications line and printing results at the same time.

- The printer will generate an interrupt every time that it completes a print operation.

- The communication line controller will generate an interrupt every time a unit of data arrives. The unit could either be a single character or a block, depending on the nature of the communications discipline.
- In any case, it is possible for a communications interrupt to occur while a printer interrupt is being processed.

Two approaches can be taken to dealing with multiple interrupts.

i. Sequential interrupt processing (disable interrupts while an interrupt is being processed)

ii. Nested interrupt processing (use a priority scheme).

i. Sequential interrupt processing

- The first is to disable interrupts while an interrupt is being processed. A disabled interrupt t simply means that the processor ignores any new interrupt request signal. If an interrupt occurs during this time, it generally remains pending and will be checked by the processor after the processor has reenabled interrupts.
- Thus, if an interrupt occurs when a user program is executing, then interrupts are disabled immediately.
- After the interrupt-handler routine completes, interrupts are reenabled before resuming the user program and the processor checks to see if additional interrupts have occurred. This approach is simple, as interrupts are handled in strict sequential order (Figure1.12a).

Drawback

1. It does not take into account relative priority or time-critical needs.
2. For example, when input arrives from the communications line, it may need to be absorbed rapidly to make room for more input.
3. If the first batch of input has not been processed before the second batch arrives, data may be lost because the buffer on the I/O device may fill and overflow.

ii. Nested interrupt processing

- A second approach is to define priorities for interrupts and to allow an interrupt of higher priority to cause a lower-priority interrupt handler to be interrupted (Figure 1.12b).

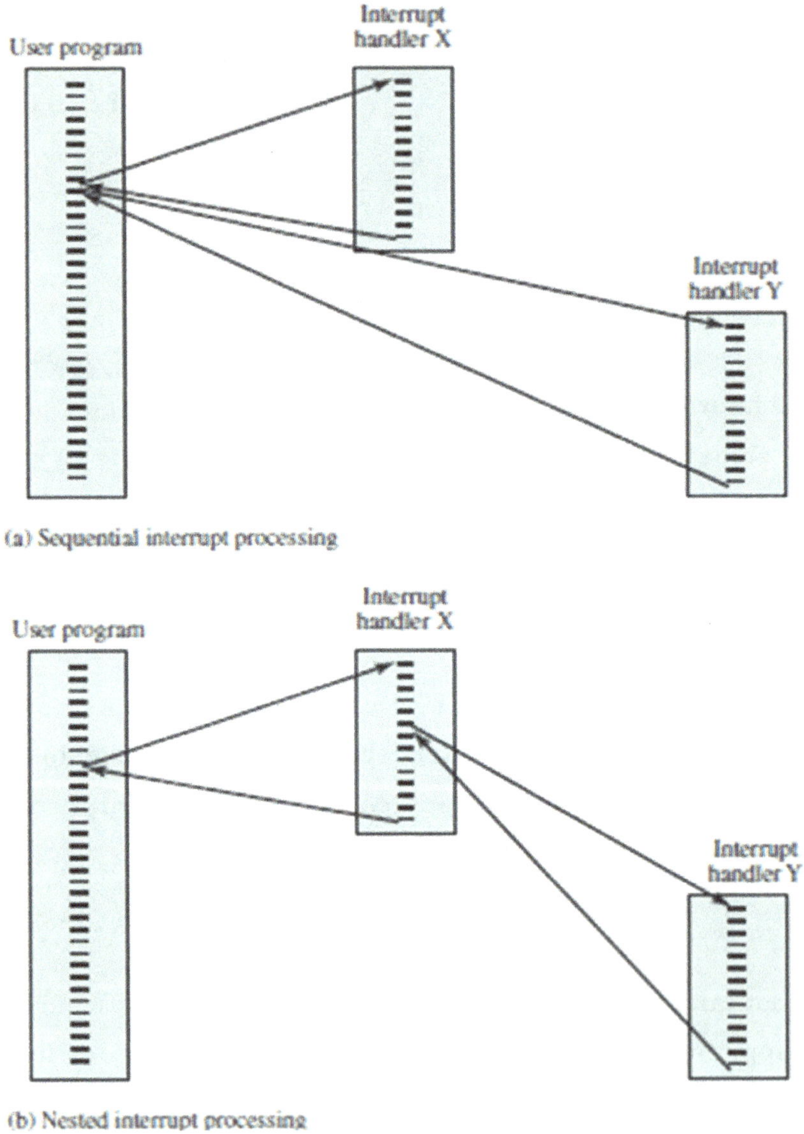

Figure 1.12 Transfer of Control with Multiple Interrupts

1.3.2 Interrupt Processing

An interrupt triggers a number of events, both in the processor hardware and in software. Figure 1.10 shows a typical sequence.

When an I/O device completes an I/O operation, the following sequence of hardware events occurs:

1. The device issues an interrupt signal to the processor.
2. The processor finishes execution of the current instruction before responding to the interrupt, as indicated in Figure 1.7.
3. The processor tests for a pending interrupt request, determines that there is one, and sends an acknowledgment signal to the device that issued the interrupt. The acknowledgment allows the device to remove its interrupt signal.
4. The processor next needs to prepare to transfer control to the interrupt routine.
   - To begin, it saves information needed to resume the current program at the point of interrupt. The minimum information required is the program status word 3 (PSW) and the location of the next instruction to be executed, which is contained in the program counter (PC). These can be pushed onto a control stack.
   - (The PSW contains status information about the currently running process, including memory usage information, condition codes, and other status information, such as an interrupt enable/disable bit and a kernel/user mode bit.)
5. The processor then loads the program counter with the entry location of the interrupt-handling routine that will respond to this interrupt.
   - Depending on the computer architecture and OS design, there may be a single program, one for each type of interrupt, or one for each device and each type of interrupt.
   - If there is more than one interrupt-handling routine, the processor must determine which one to invoke. Once the program counter has been loaded, the processor proceeds to the next instruction cycle, which begins with an instruction fetch. control is transferred to the interrupt-handler program.

6. The execution of this program results in the following operations:

   i. The program counter and PSW relating to the interrupted program have been saved on the control stack, and also the contents of the processor registers need to be saved, Figure 1.11a shows a simple example.

   ii. The interrupt handler now proceeds to process the interrupt. This includes an examination of status information relating to the I/O operation or other event that caused an interrupt.

   iii. When interrupt processing is complete, the saved register values are retrieved from the stack and restored to the registers (e.g., see Figure 1.11b).

   iv. The final act is to restore the PSW and program counter (PC) values from the stack.

It is important to save all of the state information about the interrupted program for later resumption. This is because the interrupt is not a routine called from the program.

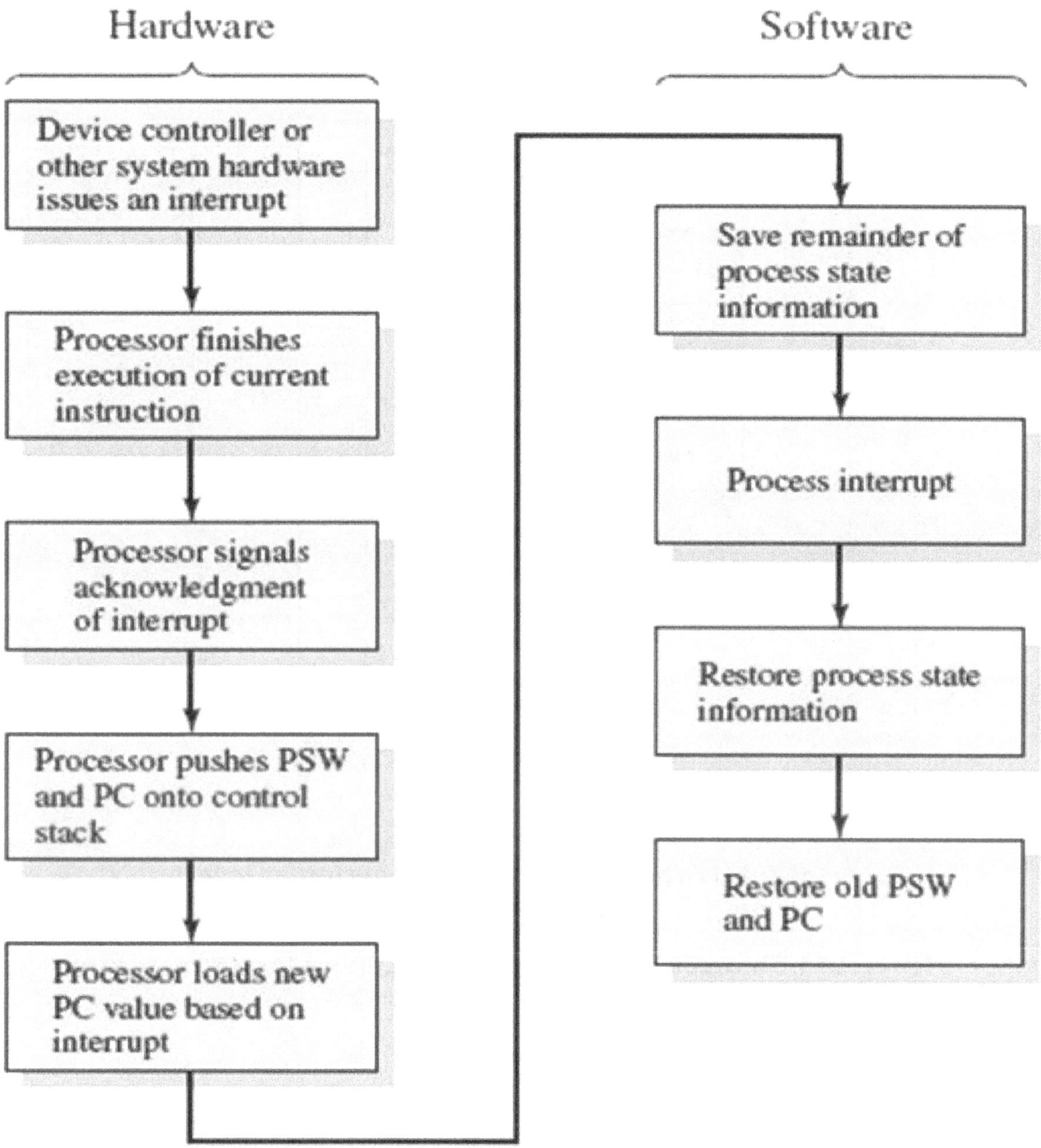

Figure 1.13 Simple Interrupt Processing

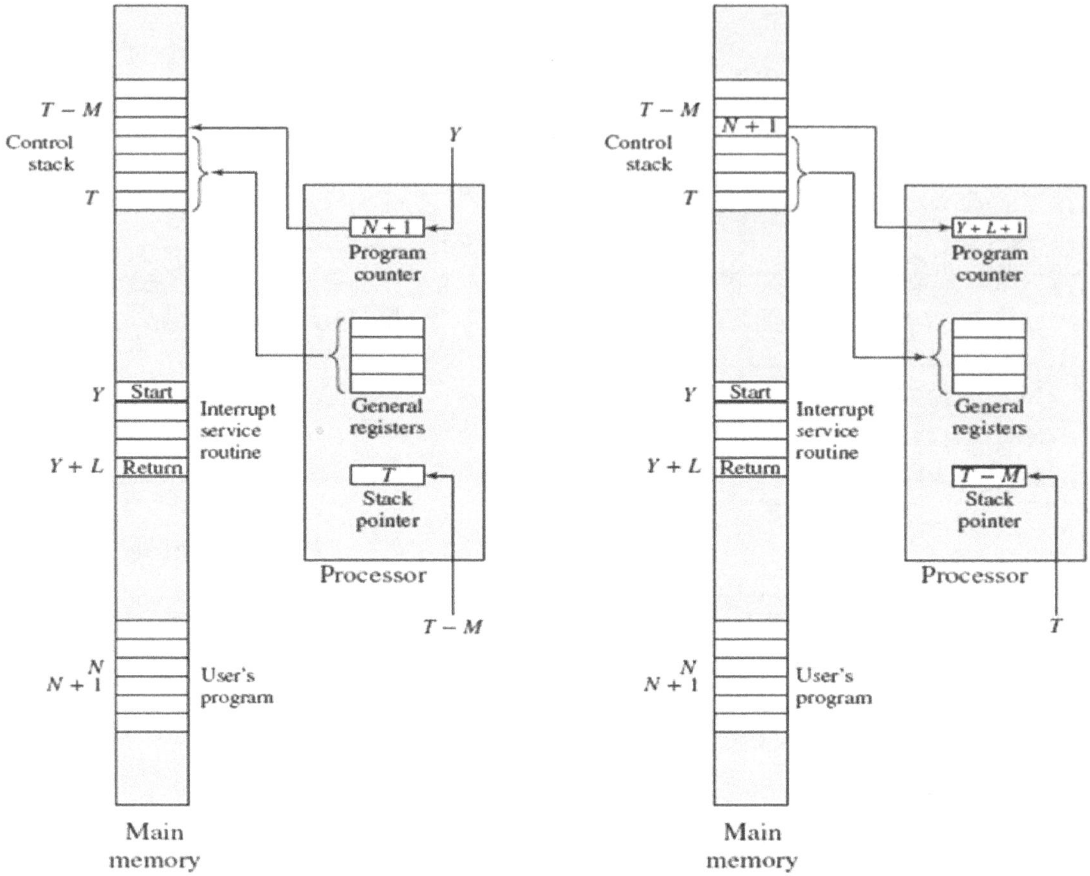

(a) Interrupt occurs after instruction at location N

(b) Return from interrupt

Figure 1.14 Changes in Memory and Registers for an Interrupt

1.4 MEMORY HIERARCHY

- Major constraints in memory
  - a. amount
  - b. speed
  - c. expense
- Memory must be able to keep up with the processor
- Cost of memory must be reasonable in relationship to the other components

A variety of technologies are used to implement memory systems, and across this spectrum of technologies, the following relationships hold:

a. Faster access time = greater cost per bit
b. Greater capacity = smaller cost per bit
c. Greater capacity = slower access speed

Memory hierarchy shown in (Figure 1.14).

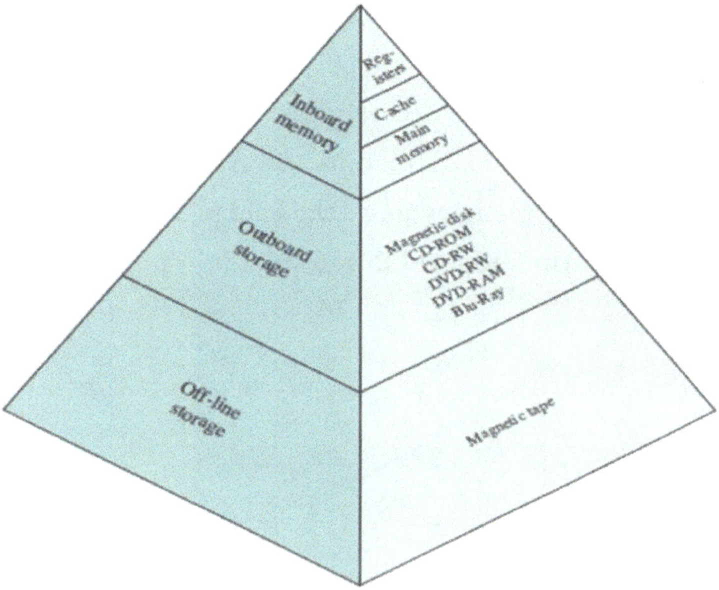

Figure. 1.15 Memory Hierarchy

As goes down in the memory hierarchy, the following has to occur:

a. Decreasing cost per bit
b. Increasing capacity
c. Increasing access time
d. Decreasing frequency of access to the memory by the processor

- Thus, smaller, more expensive, faster memories are supplemented by larger, cheaper, slower memories.
- The key to the success of this organization is the decreasing frequency of access at lower levels.

Suppose that the processor has access to two levels of memory. Level 1 contains1,000 bytes and has an access time of 0.1 μs; level 2 contains 100,000 bytes and has an access time of 1 μs.

- Assume that if a byte to be accessed is in level 1, then the processor accesses it directly. If it is in level 2, then the byte is first transferred to level 1 and then accessed by the processor. The time required for the processor to determine whether the byte is in level 1 or level 2 is ignored.
- Figure 1.15 shows the general shape of the curve that models this situation. The figure shows the average access time to a two-level memory as a function of the hit ratio H , where H is defined as the fraction of all memory accesses that are found in the faster memory (e.g., the cache), T1 is the access time to level 1, and T2 is the access time to level 2.

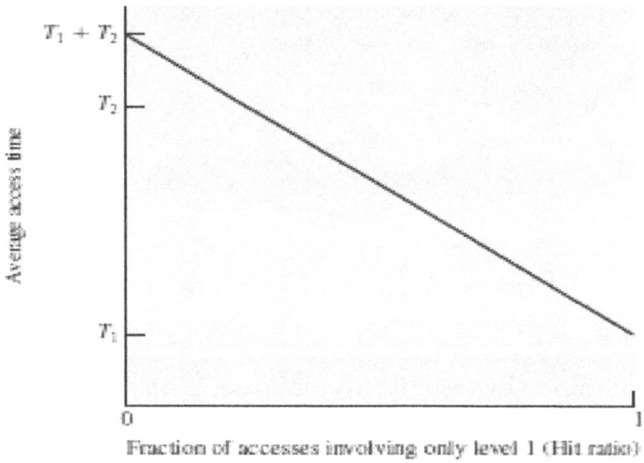

Figure 1.16 Performance of a Simple Two-Level Memory

For high percentages of level 1 access, the average total access time is much closer to that of level 1 than that of level 2.

In our example, suppose 95% of the memory accesses are found in the cache (H = 0.95) . Then the average time to access a byte can be expressed as

(0.95) (0.1 μs) _ (0.05) (0.1 μs + 1 μs) = 0.095 + 0.055= 0.15 μs

The result is close to the access time of the faster memory. So the strategy of using two memory levels works in principle, but only if conditions (a) through (d) in the preceding list applies.

By employing a variety of technologies, a spectrum of memory systems exists that satisfies conditions (a) through (c). Fortunately, condition (d) is also generally valid.

Principle of Locality

The basis of the validity of "decreasing frequency of access to the memory by the processor" is a principle known as locality of reference. This principle can be applied across more than two levels of memory.

From the memory hierarchy

1. **Main memory**
   - Main memory is the principal internal memory system of instructions refers to one or more main memory addresses.
   - Main memory is usually extended with a higher-speed, smaller cache. The cache is not usually visible to the programmer or, indeed, to the processor.
2. **External, nonvolatile memory**
   - External, nonvolatile memory is also referred to as secondary memory or auxiliary memory.
   - Data are stored more permanently on external mass storage devices, such as hard disk and removable media, such as removable disk, tape, and optical storage.
   - These are used to store program and data files, and are usually visible to the programmer only in terms of files and records, as opposed to individual bytes or words.
   - A hard disk is also used to provide an extension to main memory known as virtual memory.

Additional levels can be effectively added to the hierarchy in software.

For example

A portion of main memory can be used as a buffer to temporarily hold data that are to be read out to disk, which s is referred to as a disk cache, which improves performance in two ways:

- Disk writes are clustered. Instead of many small transfers of data, which improves disk performance and minimizes processor involvement
- Some data destined for write-out may be referenced by a program before the next dump to disk.

1.5 CACHE MEMORY

- Invisible to the OS
- Interacts with other memory management hardware
- Processor must access memory at least once per instruction cycle
- Processor execution is limited by memory cycle time
- Exploit the principle of locality with a small, fast memory

1.5.1 Motivation

On all instruction cycles, the processor accesses memory at least once, to fetch the instruction, and often one or more additional times, to fetch operands and or store results. The rate at which the processor can execute instructions is clearly limited by the memory cycle time (the time it takes to read one word from or write one word to memory).

This limitation has been a significant problem because processor speed has consistently increased more rapidly than memory access speed.

Ideally, main memory should be built with the same technology as that of the processor registers, giving memory cycle times comparable to processor cycle times which is expensive.

The solution is to exploit the principle of locality by providing a small, fast memory between the processor and main memory, namely the cache.

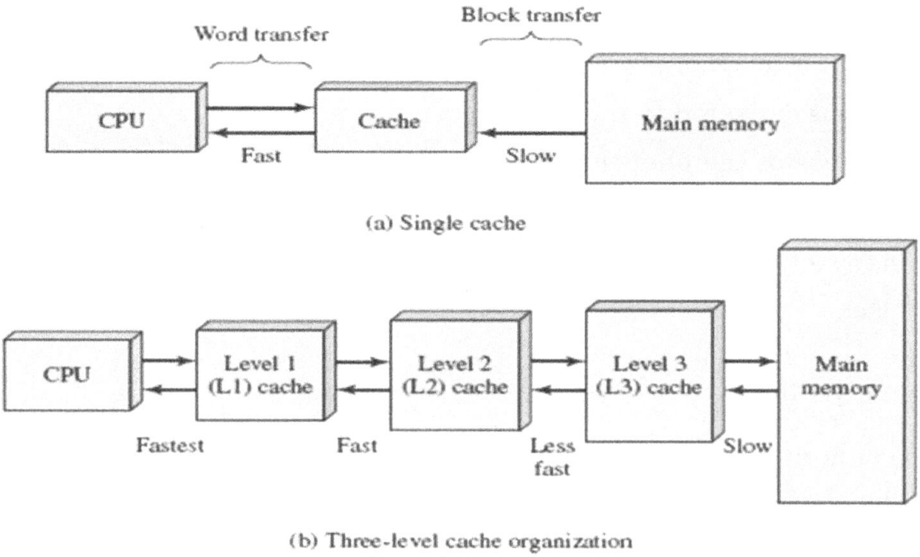

Figure 1.17 Cache and Main Memory

1.5.2 Cache Principles:

Cache memory is intended to

- Provide memory access time approaching that of the fastest memories available and
- Support a large memory size that has the price of less expensive types of semiconductor memories.

There is a relatively large and slow main memory together with a smaller, faster cache memory.

- Contains a copy of a portion of main memory
- Processor first checks cache
  - If not found, a block of memory is read into cache
- Because of locality of reference, it is likely that many of the future memory references will be to other bytes in the block

From Fig 1.16 a

- The cache contains a copy of a portion of main memory. When the processor attempts to read a byte or word of memory, a check is made to determine if the byte or word is in the cache.

- ❖ If so, the byte or word is delivered to the processor.
- ❖ If not, a block of main memory is read into the cache and then the byte or word is delivered to the processor.
- ❖ Because of the phenomenon of locality of reference, when a block of data is fetched into the cache to satisfy a single memory reference, it is likely that many of the near-future memory references will be to other bytes in the block.

From Figure 1.16b

It is the use of multi levels cache

- ➢ The L2 cache is slower and typically larger than the L1 cache, and the L3 cache is slower and typically larger than the L2 cache.

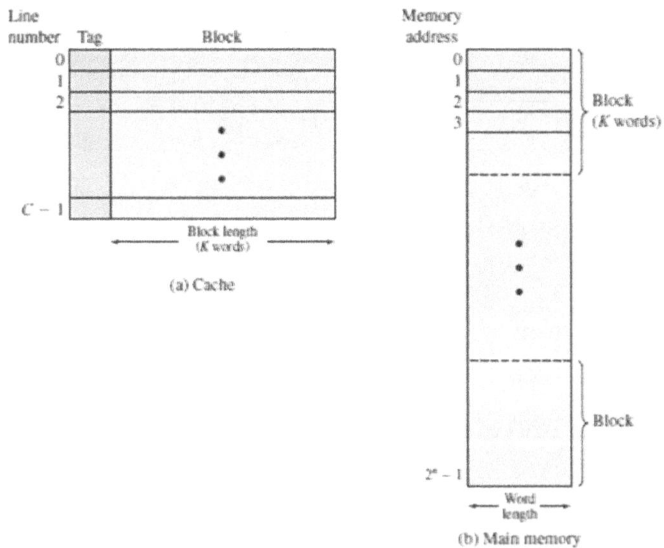

Figure 1.18 Cache/Main-Memory Structure

The structure of a cache/main memory system Figure 1.17 depicts

Main memory

- ➢ Consists of up to 2 n addressable words, with each word having a unique n –bit address.
- ➢ For mapping purposes, this memory is considered to consist of a number of fixed-length blocks of K words each. That is, there are M=2n/K blocks.

Cache

- Consists of C slots (also referred to as lines) of K words each, and the number of slots is considerably less than the number of main memory blocks (C<<M) .
- Some subset of the blocks of main memory resides in the slots of the cache.
- If a word in a block of memory that is not in the cache is read, that block is transferred to one of the slots of the cache.

Because there are more blocks than slots, an individual slot cannot be uniquely and permanently dedicated to a particular block. Therefore, each slot includes a tag that identifies which particular block is currently being stored.

The tag is usually some number of higher-order bits of the address and refers to all addresses that begin with that sequence of bits. As a simple example, suppose that we have a 6-bit address and a 2-bit tag. The tag 01 refers to the block of locations with the following addresses:

010000, 010001, 010010, 010011, 010100, 010101, 010110, 010111, 011000, 011001, 011010, 011011, 011100, 011101, 011110, 011111.

Figure illustrates the read operation.

The processor generates the address, RA, of a word to be read. If the word is contained in the cache, it is delivered to the processor. Otherwise, the block containing that word is loaded into the cache and the word is delivered to the processor.

Cache read operation Figure 1.18 illustrates

- The processor generates the address, RA, of a word to be read.
- If the word is contained in the cache, it is delivered to the processor.
- Otherwise, the block containing that word is loaded into the cache and the word is delivered to the processor.

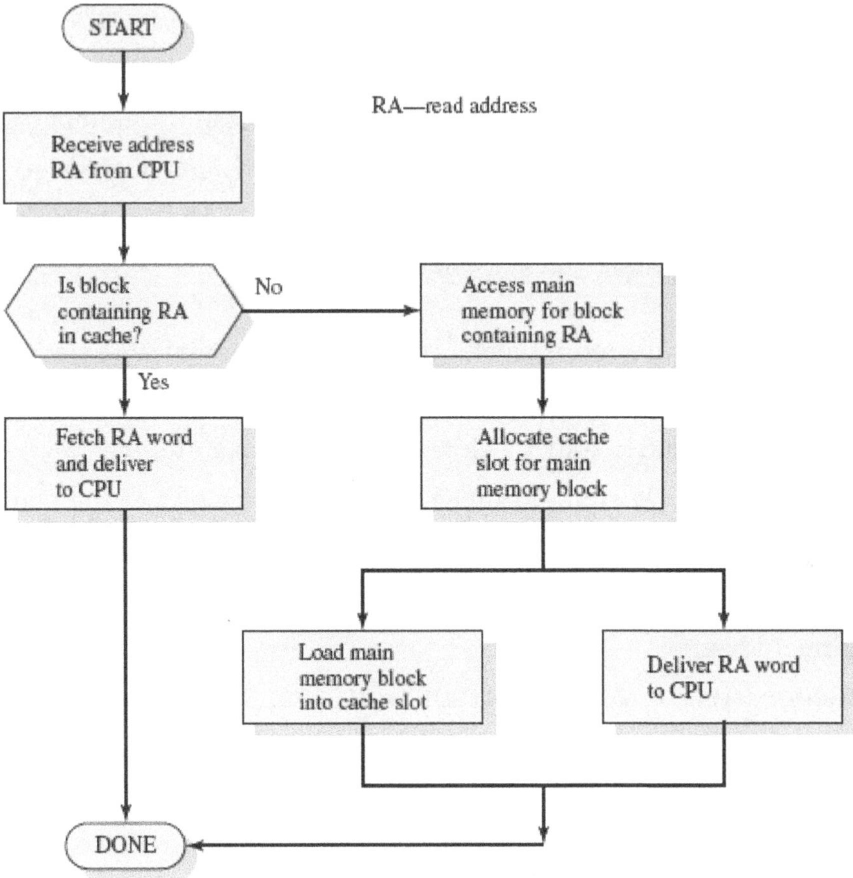

Figure 1.19 Cache Read Operation

1.5.3 Cache Design

Design issues must be addressed in dealing with virtual memory and disk cache design.

1. Cache size
2. Block size
3. Mapping function
4. Replacement algorithm
5. Write policy
6. Number of cache levels

1. Cache size

- Small caches can have a significant impact on performance.

2. Block size

- The unit of data exchanged between cache and main memory.
- As the block size increases from very small to larger sizes,
  - ➢ Hit ratio will at first increase because of the principle of locality.
  - ➢ As the block size increases, more useful data are brought into the cache.
  - ➢ The hit ratio will begin to decrease, and the probability of using the newly fetched data becomes less than the probability of reusing the data that have to be moved out of the cache to make room for the new block.

3. Mapping function

When a new block of data is read into the cache, the mapping function determines which cache location the block will occupy

Two constraints affect the design of the mapping function.

- First, when one block is read in, another may have to be replaced. Replace a block that will not be needed in the near future, and has to maximize the hit ratio.
- Second, the more flexible the mapping function, the more complex is the circuitry required to search the cache to determine if a given block is in the cache.

4. Replacement algorithm

- The replacement algorithm chooses, within the constraints of the mapping function, which block to replace when a new block is to be loaded into the cache
- Replace the block that is least likely to be needed again in the near future. Always replace the block that has been in the cache longest with no reference to it.
- This policy is referred to as the least-recently-used (LRU) algorithm.
- Hardware mechanisms are needed to identify the least-recently-used block.

5. Write policy

- If the contents of a block in the cache are altered, it is necessary to write it back to main memory before replacing it.
- The write policy dictates when the memory write operation takes place.

- At one extreme, the writing can occur every time that the block is updated.
- At the other extreme, the writing occurs only when the block is replaced.
- The latter policy minimizes memory write operations but leaves main memory in an obsolete state. This can interfere with multiple-processor operation and with direct memory access by I/O hardware modules.

6. Number of cache levels

- Finally, a common place to have multiple levels of cache
- labeled L1 (cache closest to the processor)
- L2, and in many cases a third level L3

1.6 DIRECT MEMORY ACCESS

Three techniques are possible for I/O operations

i. Programmed I/O
ii. Interrupt-DrivenI/O
iii. Direct Memory Access (DMA)

When the processor is executing a program and encounters an instruction relating to I/O, it executes that instruction by issuing a command to the appropriate I/O module.

In the case of programmed I/O

1. The I/O module performs the requested action and then sets the appropriate bits in the I/O status register but takes no action to alert the processor. It does not interrupt the processor.
2. To know the status of completion, the processor periodically checks the status of the I/O module until it finds that the operation is complete, the processor has to wait a long time for the I/O module to know the status and must repeatedly interrogate the status of the I/O module.
3. Hence, the performance level of the entire system is severely degraded.

Interrupt-driven I/O

1. It is an alternative, known as interrupt-driven I/O
2. It issues an I/O command to a module and then go on to do some other useful work.
3. The I/O module will then interrupt the processor to request service when it is ready to exchange data with the processor.
4. The processor then executes the data transfer, as before, and then resumes its former processing.
5. More efficient than the programmed I/O.

Drawbacks of both

- The I/O transfer rate is limited by the speed with which the processor can test and service a device.
- The processor is tied up in managing an I/O transfer; a number of instructions must be executed for each I/O transfer.

When large volumes of data are to be moved, a more efficient technique is required: i.e., DMA (Direct Memory Access)

Direct Memory Access (DMA)

1. When large volumes of data are to be moved, a more efficient technique is required, which is called as direct memory access (DMA).
2. The data can be moved by DMA either.
   - by a separate module on the system bus
   - or it can be incorporated into an I/O module.

Technique used

When the processor wishes to read or write a block of data, it issues a command to the DMA module, by sending to the DMA module the following information:

- Whether a read or write is requested
- The address of the I/O device involved
- The starting location in memory to read data from or write data to

- The number of words to be read or written
  - After issuing the command, the processor then continues with other work.
  - The DMA module after receiving the command from the processor
- transfers the entire block of data, one word at a time, directly to or from memory without going through the processor.
- When the transfer is complete, the DMA module sends an interrupt signal to the processor.
- Thus, the processor is involved only at the beginning and end of the transfer.

The DMA module needs to take control of the bus to transfer data to and from memory.

- When the processor needs the bus and must wait for the DMA module.
  - The processor does not save a context
  - The overall effect is to cause the processor to execute more slowly during a DMA transfer when processor access to the bus is required.
  - For a multiple-word I/O transfer, DMA is more efficient than interrupt-driven or programmed I/O.

1.7 MULITIPROCESSOR AND MULTICORE ORGANIZATION

Three most popular approaches to providing parallelism by replicating processors to improve performance and reliability, namely

1. Symmetric multiprocessors (SMPs)
2. Multicore computers
3. Clusters.

SYMMETRIC MULTIPROCESSORS (SMPS)

Definition

An SMP can be defined as a stand-alone computer system with the following characteristics:

1. There are two or more similar processors of comparable capability.

2. These processors share the same main memory and I/O facilities and are interconnected by a bus or other internal connection scheme, such that memory access time is approximately the same for each processor.
3. All processors share access to I/O devices, either through the same channels or through different
4. Channels that provide paths to the same device.
5. All processors can perform the same functions (hence the term symmetric).
6. The system is controlled by an integrated operating system that provides interaction between processors and their programs at the job, task, file, and data element levels.

An SMP organization has a number the following potential advantages over a uniprocessor organization.

- **Performance:** If the work to be done by a computer can be organized so that some portions of the work can be done in parallel, then a system with multiple processors will yield greater performance than one with a single processor of the same type.
- **Availability:** In a symmetric multiprocessor, because all processors can perform the same functions, the failure of a single processor does not halt the machine.
- **Incremental growth:** A user can enhance the performance of a system by adding an additional processor.
- **Scaling:** Vendors can offer a range of products with different price and performance characteristics based on the number of processors configured in the system.

An attractive feature of an SMP is that the existence of multiple processors is transparent to the user. The operating system takes care of scheduling of tasks on individual processors and of synchronization among processors.

ORGANIZATION of SMP

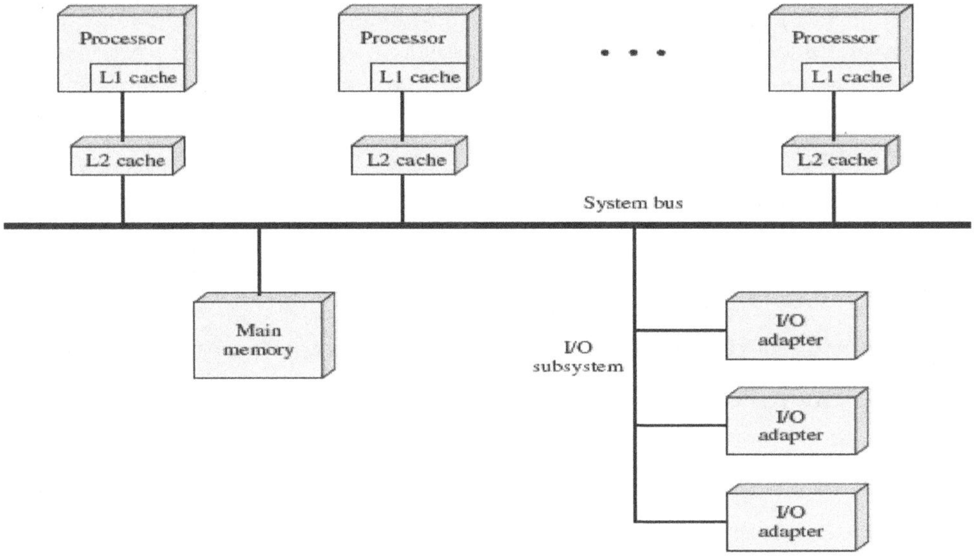

Figure 1.20 Symmetric Multiprocessor Organization

Figure 1.19 illustrates the general organization of an SMP

- There are multiple processors, each of which contains its own control unit, arithmeticlogic unit, and registers.
- Each processor has access to a shared main memory and the I/O devices through some form of interconnection mechanism; a shared bus is a common facility.
- The processors can communicate with each other through memory (messages and status information left in shared address spaces).
- Also the processors to exchange signals directly.

MULTICORE COMPUTERS

Definition

- A multicore computer, also known as a chip multiprocessor, combines two or more processors (called cores) on a single piece of silicon (called a die).
- Each core consists of all of the components of an independent processor, such as registers, ALU, pipeline hardware, and control unit, plus L1 instruction and data caches.

- Also includes L2 cache and, in some cases, L3 cache.

For decades, microprocessor systems have experienced a steady, usually exponential, increase in performance.

An example of a multicore system is the Intel Core i7:

- It includes four x86 processors, each with a dedicated L2 cache, and with a shared L3 cache (Figure 1.20).
- The Core i7 chip supports two forms of external communications to other chips.

1. The DDR3 memory controller brings the memory controller for the DDR (double data rate) main memory onto the chip.
    - The interface supports three channels that are 8 bytes wide for a total bus width of 192 bits, for an aggregate data rate of up to 32 GB/s.
2. The Quick Path Interconnect (QPI) is a point-to-point link electrical interconnect specification. It enables high-speed communications among connected processor chips. The QPI link operates at 6.4 GT/s (transfers per second).

At 16 bits per transfer, that adds up to 12.8 GB/s; and since QPI links involve dedicated bidirectional pairs, the total bandwidth is 25.6 GB/s.

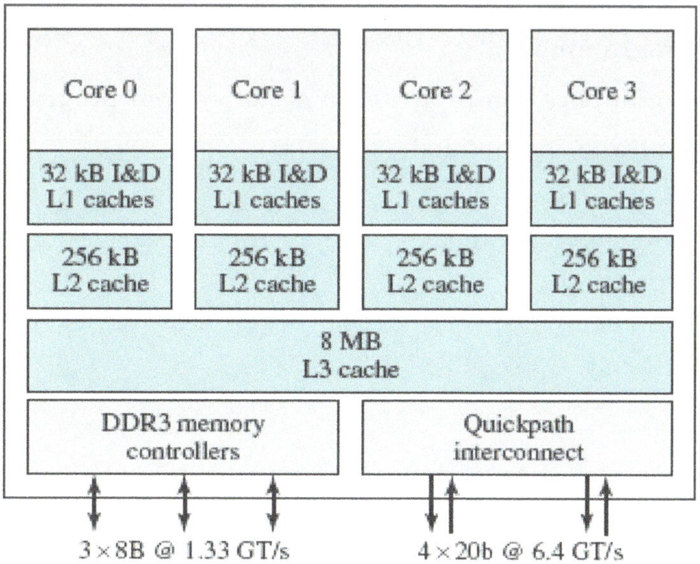

Figure 1.21 Intel Core i7 Block Diagram

1.8 OPERATING SYSTEM OVERVIEW

Definition

An OS is a program that controls the execution of application programs and acts as an interface between applications and the computer hardware.

Objectives

- **Convenience**: An OS makes a computer more convenient to use.
- **Efficiency**: An OS allows the computer system resources to be used in an efficient manner.
- **Ability to evolve**: An OS should be constructed in such a way as to permit the effective development, testing, and introduction of new system functions without interfering with service.

Three aspects of an OS

1. The Operating System as a User/Computer Interface
2. The hardware and software used can be viewed in a layered or hierarchical fashion, as depicted in Figure 1.21.
3. The end user views a computer system in terms of a set of applications.

An application can be expressed in a programming language and is developed by an application programmer.

- To develop an application program, a set of system programs called as utilities, or library programs is provided

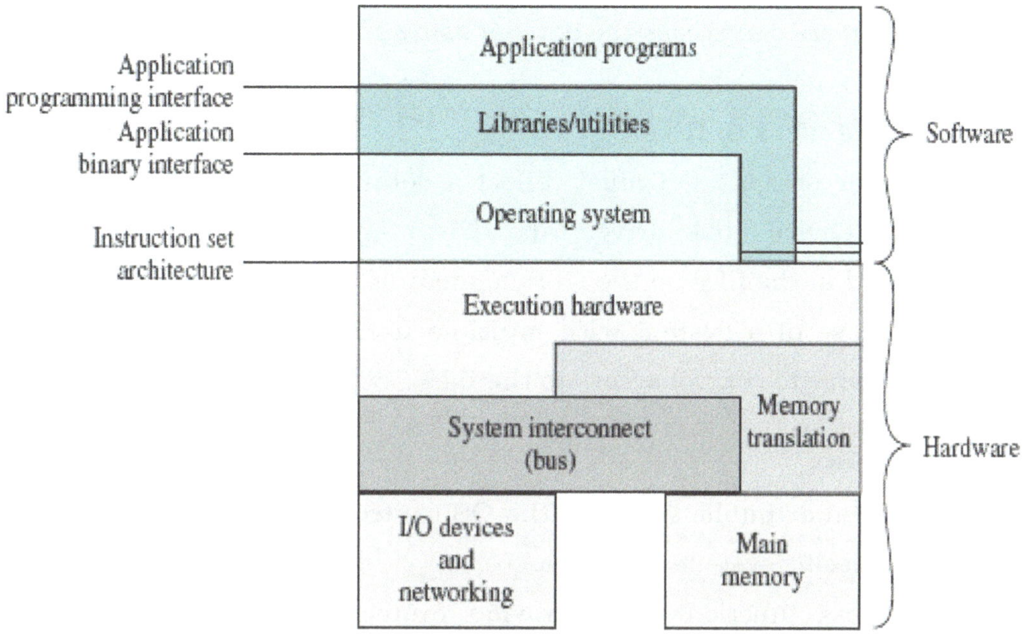

Figure 1.22 Computer Hardware and Software Structure

OS Services

1. **Program development**
   - The OS provides a variety of facilities and services, such as editors and debuggers, to assist the programmer in creating programs.
   - These services are in the form of utility programs that, while not strictly part of the core of the OS, are supplied with the OS and are referred to as application program development tools.

2. **Program execution**
   - Instructions and data must be loaded into main memory, I/O devices and files must be initialized, and other resources must be prepared.
   - The OS handles these scheduling duties for the user.

3. **Access to I/O devices**
   - Each I/O device requires its own peculiar set of instructions or control signals for operation.

- The OS provides a uniform interface that hides these details so that programmers can access the devices using simple reads and writes.

4. **Controlled access to files**
   - For file access, the OS must reflect a detailed understanding of the nature of the I/O device (disk drive, tape drive) and also the structure of the data contained in the files on the storage medium.
   - In the case of a system with multiple users, the OS may provide protection mechanisms to control access to the files.

5. **System access**
   - For shared or public systems, the OS controls access to the system as a whole and to specific system resources.
   - The access function must provide protection of resources and data from unauthorized users and must resolve conflicts for resource contention.

6. **Error detection and response**
   - A variety of errors can occur while a computer system is running, which includes
   - Internal and external hardware errors, such as a memory error, or a device failure or malfunction and various software errors, such as division by zero, attempt to access
   - In each case, the OS must provide a response that clears the error condition with the least impact on running applications.

7. **Accounting**
   - A OS will collect usage statistics for various resources and monitor performance parameters such as response time for future enhancements

Key Interfaces in a Computer System:

1. **Instruction set architecture (ISA)**
   - The ISA defines the collection of machine language instructions that a computer can follow.

- Application programs and utilities can access the ISA directly. For these programs, a subset of the instruction list is available (user ISA).
- The OS has access to additional machine language instructions that deal with managing system resources (system ISA).

2. Application binary interface (ABI)

- The ABI defines a standard for binary portability across programs.
- The ABI defines the system call interface to the operating system and the hardware resources and services available in a system through the user ISA.

3. Application programming interface (API)

- The API gives a program access to the hardware resources and services available in a system through the user ISA supplemented with high-level language (HLL) library calls.
- Any system calls are usually performed through libraries.
- Using an API enables application software to be ported easily, through recompilation, to other systems that support the same API.

The Operating System as Resource Manager:

A computer is a set of resources for the movement, storage, and processing of data and for the control of these functions.

The OS is responsible for managing resources:

1. The OS functions, is a program or suite of programs executed by the processor.
2. The OS frequently relinquishes control and must depend on the processor to allow it to regain control.

Figure 1.22 shows the main resources that are managed by the OS.

- A portion of the OS is in main memory, which includes kernel or nucleus and contains the most frequently used functions in the OS at a given time, other portions of the OS currently in use.
- The remainder of main memory contains user programs and data.

- The processor itself is a resource, and the OS must determine how much processor time is given to the execution of a particular user program.

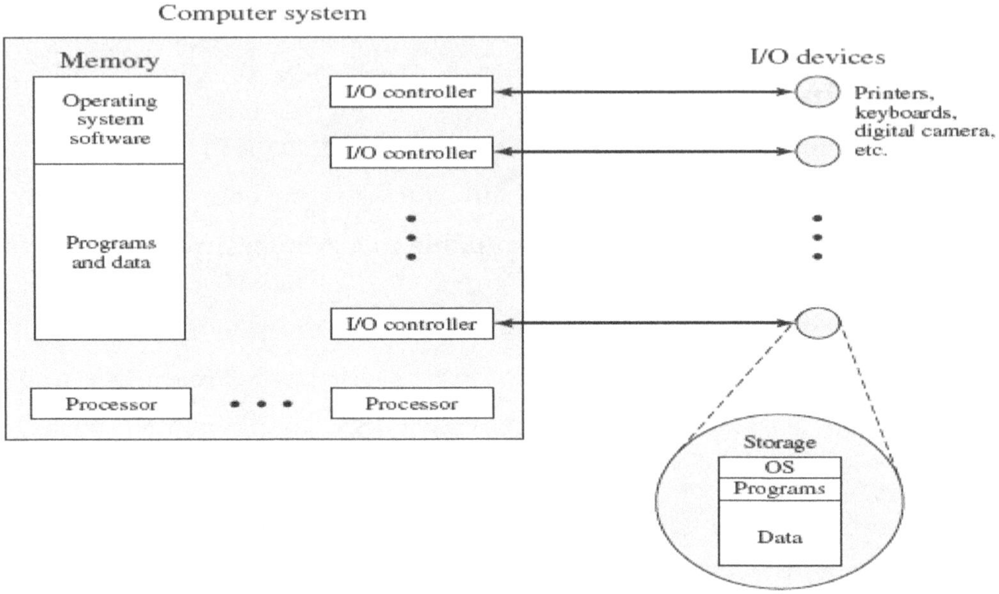

Figure 1.23 The Operating System as Resource Manager

Ease of Evolution of an Operating System:

A major OS will evolve over time for a number of reasons.

1. **Hardware upgrades plus new types of hardware**

   For example

   - Early versions of UNIX and the Macintosh OS were run on processors without paging hardware. Subsequent versions of these operating systems were modified to exploit paging capabilities.

2. **New services**

   - In response to user demand or in response to the needs of system managers, the OS expands to offer new services.
   - For example, if it is found to be difficult to maintain good performance for users with existing tools, new measurement and control tools may be added to the OS.

3. **Fixes**
   - Any OS has faults, and faults and are discovered over the course of time and fixes are made.

1.9 THE EVOLUTION OF OPERATING SYSTEMS

1. Serial Processing
2. Simple Batch Systems
3. Multiprogrammed Batch systems
4. Time sharing systems

1. **Serial Processing :** Earliest Computers (From the late 1940s to the mid-1950s)
   - No operating system
     - ➢ programmers interacted directly with the computer hardware
   - Computers ran from a console with display lights, toggle switches, some form of input device, and a printer
   - Users have access to the computer in "series"

These early systems presented two main problems

- **Scheduling**
  - ➢ Most installations used a hardcopy sign-up sheet to reserve computer time
    - ❖ Time allocations could run short or long, resulting in wasted computer time
- **Setup time**
  - ➢ A considerable amount of time was spent just on setting up the program to run

2. **Simple Batch Systems**
   - Early computers were very expensive
     - ➢ Important to maximize processor utilization
   - Monitor (The central idea behind the simple batch-processing scheme is the use of a piece of software known as the monitor)
     - ➢ User no longer has direct access to processor

- ➢ Job is submitted to computer operator who batches them together and places them on an input device
- ➢ Program branches back to the monitor when finished

Working scheme with two points of view

i. Monitor point of view
ii. Processor point of view

i. Monitor point of view

- Monitor controls the sequence of events
- Resident Monitor is software always in memory (available for execution)
- Monitor reads in job and gives control
- When job completes, Job returns control to monitor

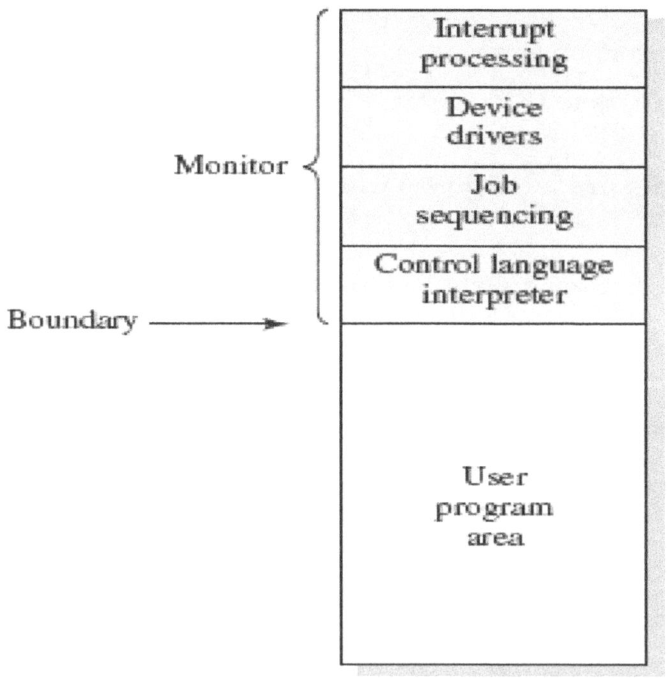

Figure 1.24 Memory Layout for a Resident Monitor

ii. **Processor point of view:**

- Processor executes instruction from the memory containing the monitor
- Executes the instructions in the user program until it encounters an ending or error condition
- "control is passed to a job" means processor is fetching and executing instructions in a user program
- "control is returned to the monitor" means that the processor is fetching and executing instructions from the monitor program.

Job Control Language (JCL)

- Special type of programming language used to provide instructions to the monitor to improve job set up time
- what compiler to use
- what data to use

A simple example

All FORTRAN instructions and data are on a separate punched card or a separate record on tape.

The monitor performs a scheduling function. The monitor, or batch OS, is simply a computer program. It relies on the ability of the processor to fetch instructions from various portions of main memory to alternately seize and relinquish control.

Hardware features:

- **Memory protection for monitor**
  - while the user program is executing, it must not alter the memory area containing the monitor
- **Timer**
  - prevents a job from monopolizing the system
- **Privileged instructions**
  - can only be executed by the monitor
- **Interrupts**
  - gives OS more flexibility in controlling user programs

Modes of Operation:

i. User mode

ii. Kernel mode

- **User Mode**
  - user program executes in user mode
  - certain areas of memory are protected from user access
  - certain instructions may not be executed

- **Kernel Mode**
  - monitor executes in kernel mode
  - privileged instructions may be executed
  - protected areas of memory may be accessed

Simple Batch System Overhead

- Processor time alternates between execution of user programs and execution of the monitor
- **Sacrifices**
  - Some main memory is now given over to the monitor
  - Some processor time is consumed by the monitor
- Most of the time the processor is in idle state.

Multiprogrammed Batch systems:

System Utilization Examples:

Read one record from file	15μs
Execute 100 instructions	1 μs
Write one record to file	15 μs
Total	31 μs

Percent CPU Utilization= 1/31 = 0.032 = 3.2%

- **Processor is often idle**
  - ➢ Even with automatic job sequencing
  - ➢ I/O devices are slow compared to processor.

Uniprogramming:

➢ The processor spends a certain amount of time executing, until it reaches an I/O instruction; it must then wait until that I/O instruction concludes before proceeding

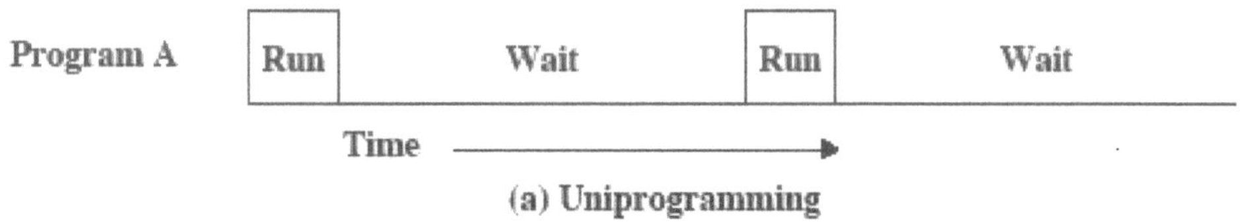

Figure 1.25 Uniprogamming

Multiprogramming:

➢ There must be enough memory to hold the OS (resident monitor) and one user program

➢ When one job needs to wait for I/O, the processor can switch to the other job, which is likely not waiting for I/O.

➢ Multiprogramming

- Also known as multitasking
- Memory is expanded to hold three, four, or more programs and switch among all of them

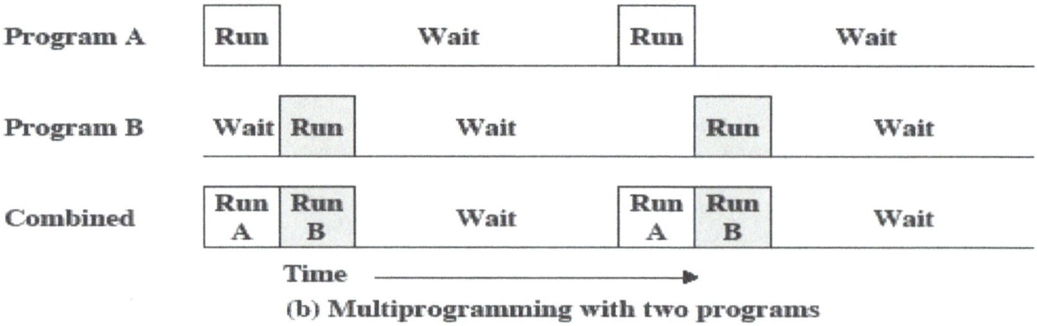

Figure 1.26 Multiprogramming with two programs

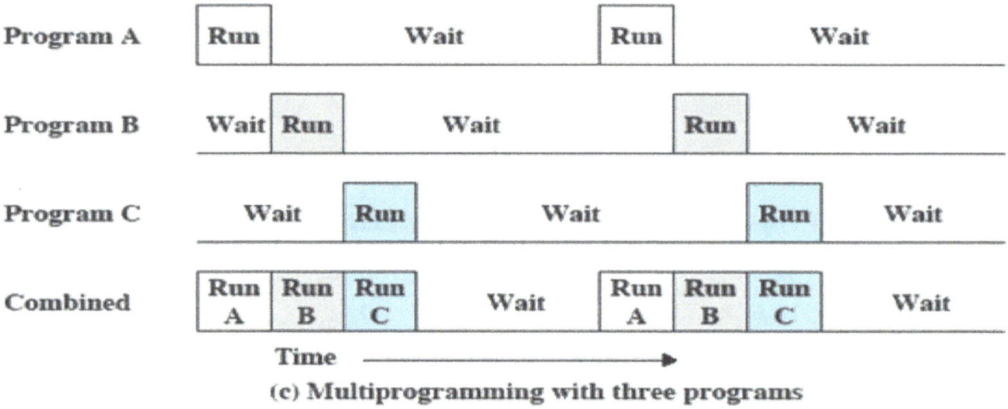

Figure 1.27 Multiprogramming with three programs

Multiprogramming example

Table 1.1 Sample Program Execution Attributes

	JOB1	JOB2	JOB3
Type of job	Heavy compute	Heavy I/O	Heavy I/O
Duration	5 min	15 min	10 min
Memory required	50 M	100 M	75 M
Need disk?	No	No	Yes
Need terminal?	No	Yes	No
Need printer?	No	No	Yes

Effects on Resource Utilization:

Table 1.2 Effects of Multiprogramming on Resource Utilization

	Uniprogramming	Multiprogramming
Processor use	20%	40%
Memory use	33%	67%
Disk use	33%	67%
Printer use	33%	67%
Elapsed time	30 min	15 min
Throughput	6 jobs/hr	12 jobs/hr
Mean response time	18 min	10 min

Time Sharing Systems

- Can be used to handle multiple interactive jobs
- Processor time is shared among multiple users
- Multiple users simultaneously access the system through terminals, with the OS interleaving the execution of each user program in a short burst or quantum of computation

Batch Multiprogramming vs. Time Sharing

Table 1.3 Batch Multiprogramming versus Time Sharing

	Batch Multiprogramming	Time Sharing
Principal objective	Maximize processor use	Minimize response time
Source of directives to operating system	Job control language commands provided with the job	Commands entered at the terminal

Compatible Time-Sharing Systems (CTTS):

CTSS

- One of the first time-sharing operating systems
- Developed at MIT by a group known as Project MAC
- Ran on a computer with 32,000 36-bit words of main memory, with the resident monitor consuming 5000 of that
- To simplify both the monitor and memory management a program was always loaded to start at the location of the 5000th word.

Time Slicing:

A system clock generated interrupts at a rate of approximately one every 0.2 seconds. At each clock interrupt, the OS regained control and could assign the processor to another user. This technique is known as time slicing

- System clock generates interrupts at a rate of approximately one every 0.2 seconds
- At each interrupt OS regained control and could assign processor to another user
- At regular time intervals the current user would be preempted and another user loaded in
- Old user programs and data were written out to disk
- Old user program code and data were restored in main memory when that program was next given a turn

CTSS Operation:

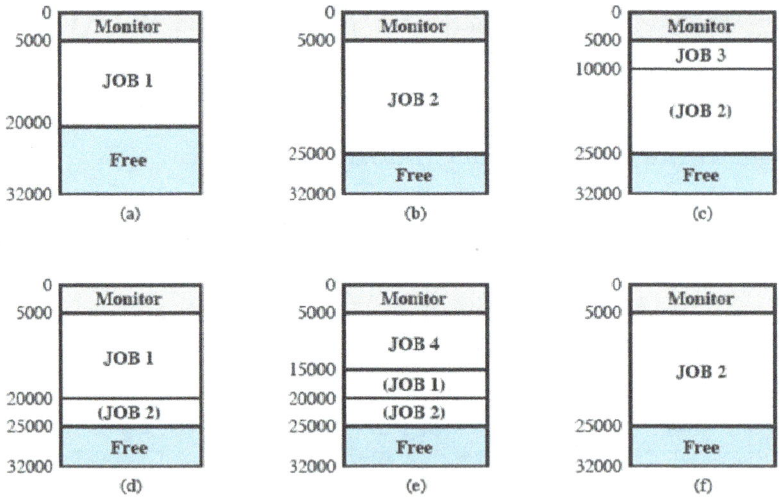

Figure 1.28 CTSS Operation

1.10 COMPUTER-SYSTEM ORGANIZATION

1.10.1 Introduction: Operating System

- A program that acts as an intermediary between a user of a computer and the computer hardware

- **Operating system goals:**
  - ➢ Execute user programs and make solving user problems easier
  - ➢ Make the computer system convenient to use
  - ➢ Use the computer hardware in an efficient manner

Computer System Structure

- **Computer system can be divided into four components:**
  - ➢ Hardware – provides basic computing resources
    - ❖ CPU, memory, I/O devices
  - ➢ Operating system
    - ❖ Controls and coordinates use of hardware among various applications and users
  - ➢ Application programs – define the ways in which the system resources are used to solve the computing problems of the users
    - ❖ Word processors, compilers, web browsers, database systems, video games
  - ➢ Users
    - ❖ People, machines, other computers.

Four Components of a Computer System:

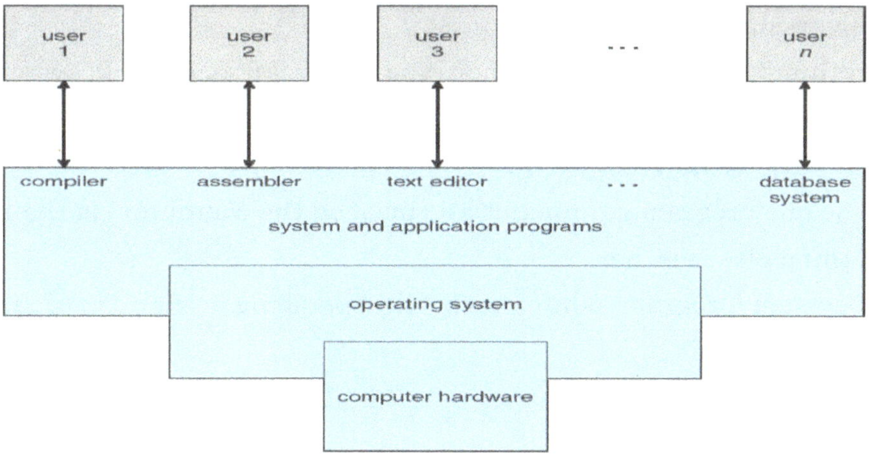

Figure 1.29 Four Components of a Computer system

What Operating Systems Do:

- Depends on the point of view
- Users want convenience, ease of use and good performance
- Don't care about resource utilization
- But shared computer such as mainframe or minicomputer must keep all users happy
    - Users of dedicate systems such as workstations have dedicated resources but frequently use shared resources from servers
- Handheld computers are resource poor, optimized for usability and battery life
- Some computers have little or no user interface, such as embedded computers in devices and automobiles

Operating System Definition:

- OS is a resource allocator
    - Manages all resources
    - Decides between conflicting requests for efficient and fair resource use
- OS is a control program
    - Controls execution of programs to prevent errors and improper use of the computer
- No universally accepted definition
- "Everything a vendor ships when you order an operating system" is a good approximation
    - But varies wildly
    - "The one program running at all times on the computer" is the kernel.
- Everything else is either
    - A system program (ships with the operating system) , or an application program.

Computer Startup:

- bootstrap program is loaded at power-up or reboot
    - Typically stored in ROM or EPROM, generally known as firmware
    - Initializes all aspects of system
    - Loads operating system kernel and starts execution

1.10.2 Computer-system operation

- One or more CPUs, device controllers connect through common bus providing access to shared memory
- Concurrent execution of CPUs and devices competing for memory cycles.

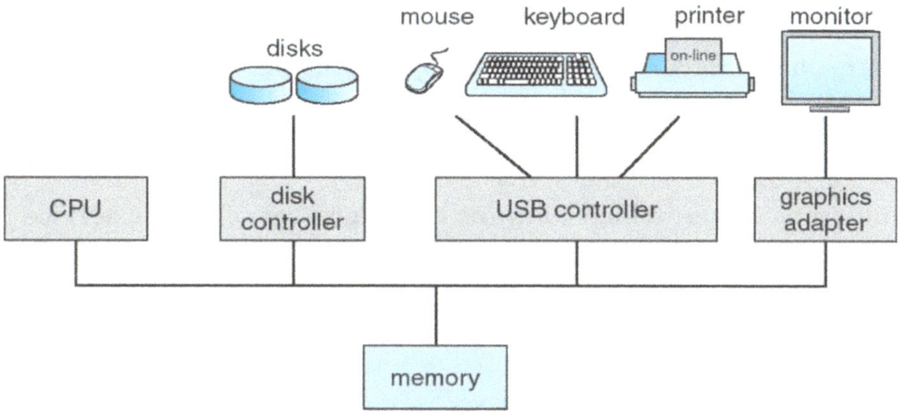

Figure 1.30 Modern computer system

- I/O devices and the CPU can execute concurrently
- Each device controller is in charge of a particular device type
- Each device controller has a local buffer
- CPU moves data from/to main memory to/from local buffers
- I/O is from the device to local buffer of controller
- Device controller informs CPU that it has finished its operation by causing an interrupt.

Common Functions of Interrupts

- Once the kernel is loaded and executing, it can start providing services to the system and its users.
- Some services are provided outside of the kernel, by system programs that are loaded into memory at boot time to become system processes, or system daemons that run the entire time the kernel is running. On UNIX, the first system process is "init," and it starts many other daemons.

- Once this phase is complete, the system is fully booted, and the system waits for some event to occur. The occurrence of an event is usually signaled by an interrupt from either the hardware or the software.
- Hardware may trigger an interrupt at any time by sending a signal to the CPU, usually by way of the system bus.
- Software may trigger an interrupt by executing a special operation called a system call (also called a monitor call).
- Interrupt transfers control to the interrupt service routine generally, through the interrupt vector, which contains the addresses of all the service routines
- Interrupt architecture must save the address of the interrupted instruction
- A trap or exception is a software-generated interrupt caused either by an error or a user request
- An operating system is interrupt driven.

When the CPU is interrupted, it stops what it is doing and immediately transfers execution to a fixed location. The fixed location usually contains the starting address where the service routine for the interrupt is located. The interrupt service routine executes; on completion, the CPU resumes the interrupted computation

Interrupt Handling

- The operating system preserves the state of the CPU by storing registers and the program counter
- Determines which type of interrupt has occurred:
    - polling
    - vectored interrupt system
- Separate segments of code determine what action should be taken for each type of interrupt

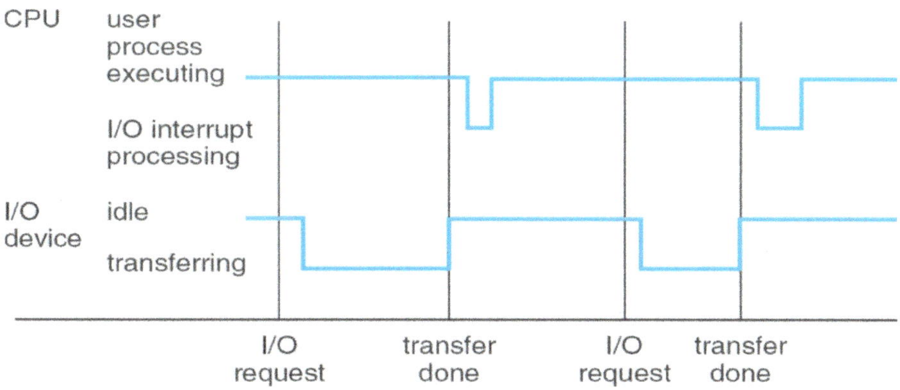

Figure 1.31 Interrupt timeline for a single process doing output

1.10.3 Storage Structure

- The CPU can load instructions only from memory.
- General-purpose computers run most of their programs from rewritable memory, called main memory (also called random-access memory, or RAM).
- Other type of memory is read-only memory, ROM and electrically erasable programmable read-only memory, EEPROM – used to store bootstrap program
- Main memory – only large storage media that the CPU can access directly
  - Random access
  - Typically volatile
  - dynamic random-access memory(DRAM)
- Secondary storage – extension of main memory that provides large nonvolatile storage capacity
- Hard disks – rigid metal or glass platters covered with magnetic recording material
  - Disk surface is logically divided into tracks, which are subdivided into sectors
  - The disk controller determines the logical interaction between the device and the computer
- Solid-state disks – faster than hard disks, nonvolatile
  - Various technologies
  - Becoming more popular.

Storage hierarchy

- Storage systems organized in hierarchy
  - Speed
  - Cost
  - Volatility
- Caching – copying information into faster storage system; main memory can be viewed as a cache for secondary storage
- Device Driver for each device controller to manage I/O
  - Provides uniform interface between controller and kernel.

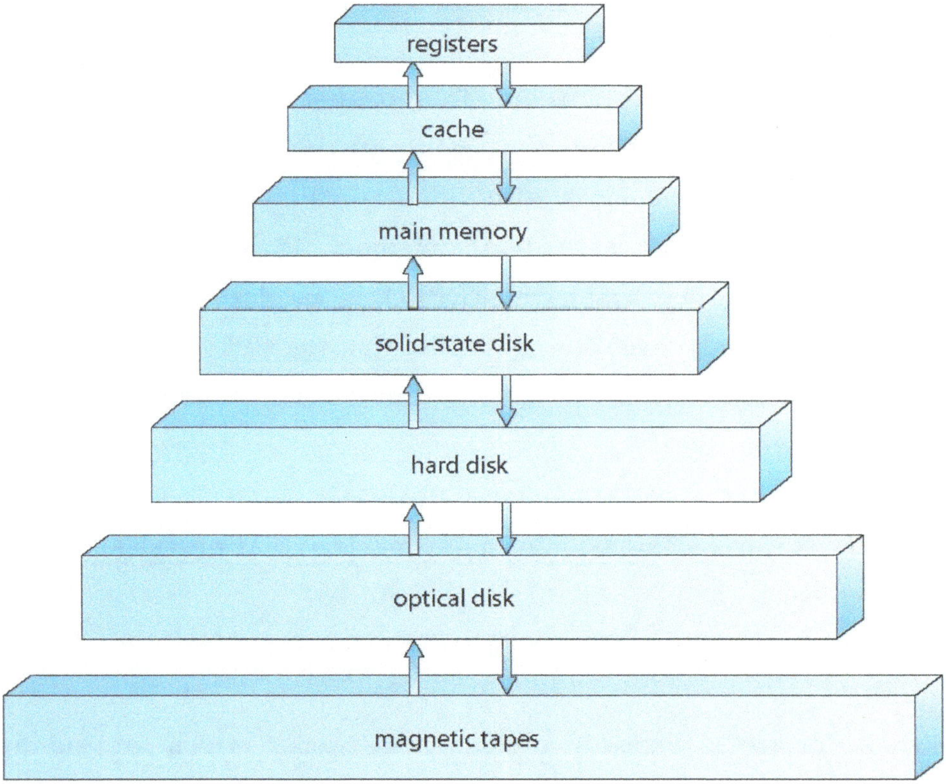

Figure 1.32 Storage device hierarchy

Caching

- Important principle, performed at many levels in a computer (in hardware, operating system, software)
- Information in use copied from slower to faster storage temporarily

- Faster storage (cache) checked first to determine if information is there
  - If it is, information used directly from the cache (fast)
  - If not, data copied to cache and used there
- Cache smaller than storage being cached
  - Cache management important design problem
  - Cache size and replacement policy.

Direct Memory Access Structure:

- Used for high-speed I/O devices able to transmit information at close to memory speeds
- Device controller transfers blocks of data from buffer storage directly to main memory without CPU intervention
- Only one interrupt is generated per block, rather than the one interrupt per byte.

I/O Structure

- After I/O starts, control returns to user program only upon I/O completion
  - Wait instruction idles the CPU until the next interrupt
  - Wait loop (contention for memory access)
  - At most one I/O request is outstanding at a time, no simultaneous I/O processing
- After I/O starts, control returns to user program without waiting for I/O completion
  - System call – request to the OS to allow user to wait for I/O completion
  - Device-status table contains entry for each I/O device indicating its type, address, and state
  - OS indexes into I/O device table to determine device status and to modify table entry to include interrupt

1.11 OPERATING-SYSTEM STRUCTURE

One of the most important aspects of operating systems is the ability to multi program. Multiprogramming increases CPU utilization by organizing jobs (code and data) so that the CPU always has one to execute. Multiprogrammed systems provide an environment in which the various

system resources (for example, CPU, memory, and peripheral devices) are utilized effectively, but they do not provide for user interaction with the computer system.

Multiprogramming (Batch system) needed for efficiency

- Single user cannot keep CPU and I/O devices busy at all times
- Multiprogramming organizes jobs (code and data) so CPU always has one to execute
- In general, main memory is too small to accommodate all jobs; the jobs are kept initially on the disk in the job pool. This pool consists of all processes residing on disk awaiting allocation of main memory
- A subset of total jobs in system is kept in memory
- One job selected and run via job scheduling
- When it has to wait (for I/O for example), OS switches to another job
- In a non-multiprogrammed system, the CPU sits idle.

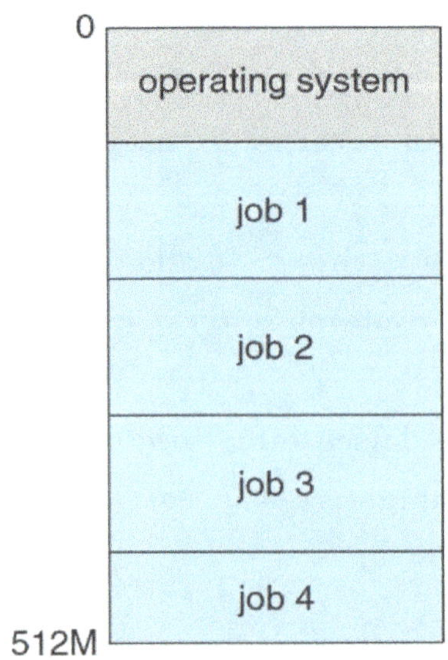

Figure 1.33 Memory layout for a multiprogramming system.

Timesharing (multitasking) is logical extension in which CPU switches jobs so frequently that users can interact with each job while it is running, creating interactive computing

- Response time should be < 1 second
- It uses CPU scheduling and multiprogramming to provide each user with a small portion of a time-shared computer.
- Each user has at least one program executing in memory ⇨process
- If several jobs are ready to be brought into memory, and if there is not enough room for all of them, then the system must choose among them. Making this decision involves job scheduling or CPU scheduling
- If processes don't fit in memory, swapping moves them in and out to run
- Virtual memory allows execution of processes not completely in memory-

1.12 OPERATING-SYSTEM OPERATIONS

Modern operating systems are interrupt driven. If there are no processes to execute, no I/O devices to service, and no users to whom to respond, an operating system will sit quietly, waiting for something to happen.

Events are almost always signaled by the occurrence of an interrupt or a trap.

Operating-System Operations

1. **Interrupt driven (**hardware and software)
   - Hardware interrupt by one of the devices
   - Software interrupt (exception or trap):
     - A trap (or an exception) is a software-generated interrupt caused either by an error (for example, division by zero or invalid memory access) or by a specific request from a user program that an operating-system service be performed.
     - Request for operating system service.
     - For each type of interrupt, separate segments of code in the operating system determine what action should be taken.
     - Other process problems include infinite loop, processes modifying each other or the operating system.

2. Dual-and multimode operation allows OS to protect itself and other system components

- To ensure the proper execution of the operating system
    - operating-system code and user defined code need to be differentiated properly
- **Dual-mode** operation allows OS to protect itself and other system components
    - User mode and kernel mode
    - Mode bit provided by hardware
        - ✓ Provides ability to distinguish when system is running user code or kernel code
        - ✓ Some instructions designated as privileged, only executable in kernel mode
        - ✓ System call changes mode to kernel, return from call resets it to user
        - Increasingly CPUs support multi-mode operations

    i.e. virtual machine manager (VMM) mode for guest VMs.

3. Increasingly CPUs support multi-mode operations

i.e. virtual machine manager (VMM) mode for guest VMs

Transition from User to Kernel Mode

- Timer to prevent infinite loop / process hogging resources
- Timer is set to interrupt the computer after some time period
- Keep a counter that is decremented by the physical clock.
- Operating system set the counter (privileged instruction)
- When counter zero generate an interrupt
- Set up before scheduling process to regain control or terminate program that exceeds allotted time

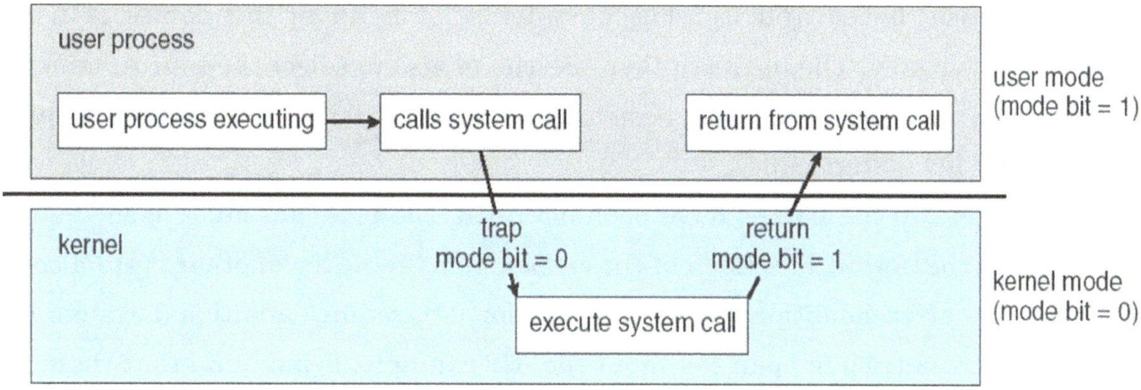

Figure 1.34 Transition from user to kernel mode.

1.13 SYSTEM CALLS

1.13.1 Definition:

System calls provide an interface to the services made available by an operating system.

- Programming interface to the services provided by the OS
- Typically written in a high-level language (C or C++)
- Mostly accessed by programs via a high-level Application Programming Interface (API) rather than direct system call use

Example to illustrate how system calls are used:

- Writing a simple program
- To read data from one file and
- Copy them to another file.

The first input that the program will need is the names of the two files: the input file and the output file.

These names can be specified in many ways, depending on the operating-system design. One approach is for the program to ask the user for the names.

In an interactive system, this approach will require a sequence of system calls, first to write a prompting message on the screen and then to read from the keyboard the characters that define the two files.

On mouse- based and icon-based systems, a menu of file names is usually displayed in a window. The user can then use the mouse to select the source name, and a window can be opened for the destination name to be specified. This sequence requires many I/O system calls.

Once the two file names have been obtained, the program must open the input file and create the output file. Each of these operations requires another system call.

Possible error conditions for each operation can require additional system calls. When the program tries to open the input file, for example, it may find that there is no file of that name or that the file is protected against access. In these cases, the program should print a message on the console (another sequence of system calls) and then terminate abnormally (another system call).

If the input file exists, then we must create a new output file. We may find that there is already an output file with the same name. This situation may cause the program to abort (a system call), or we may delete the existing file (another system call) and create a new one (yet another system call).

Finally, after the entire file is copied, the program may close both files (another system call), write a message to the console or window (more system calls), and finally terminate normally (the final system call). This system-call sequence is shown in Figure.

Typically, application developers design programs according to an Application Programming Interface (API). The API specifies a set of functions that are available to an application programmer, including the parameters that are passed to each function and the return values the programmer can expect.

Three most common APIs are

- Win32 API for Windows,
- POSIX API for POSIX-based systems (including virtually all versions of UNIX, Linux,
- Mac OS X), and Java API for the Java virtual machine (JVM)
- System call sequence to copy the contents of one file to another file

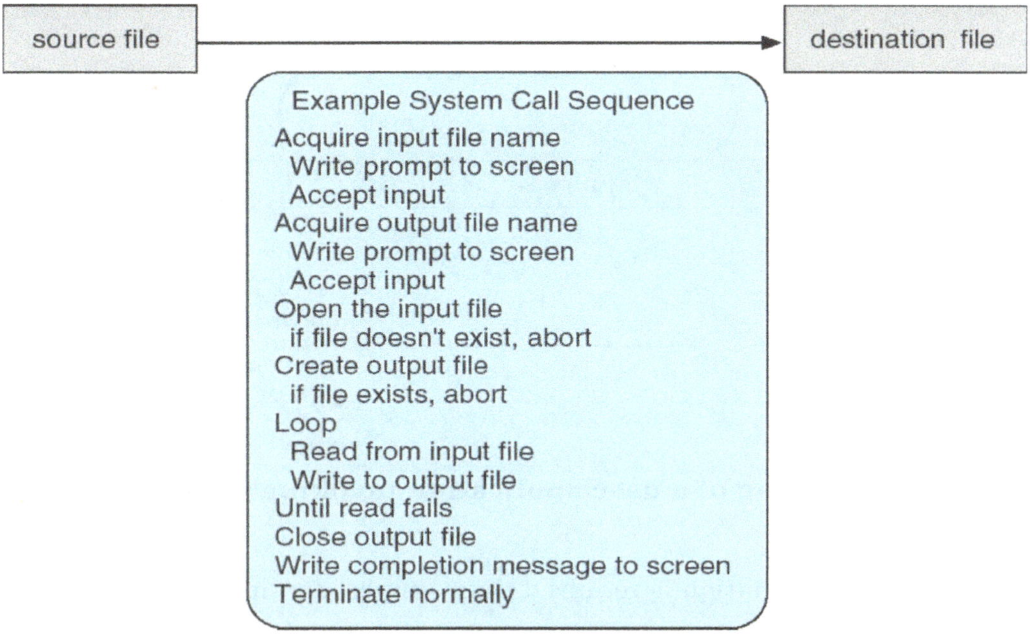

Figure 1.35 Example of how system calls are used

1.13.2 System Call Implementation

- Typically, a number associated with each system call
  - ✓ System-call interface maintains a table indexed according to these numbers
- The system call interface invokes the intended system call in OS kernel and returns status of the system call and any return values
- The caller need know nothing about how the system call is implemented
  - ✓ Just needs to obey API and understand what OS will do as a result call
  - ✓ Most details of OS interface hidden from programmer by API
    - ❖ Managed by run-time support library (set of functions built into libraries included with compiler).

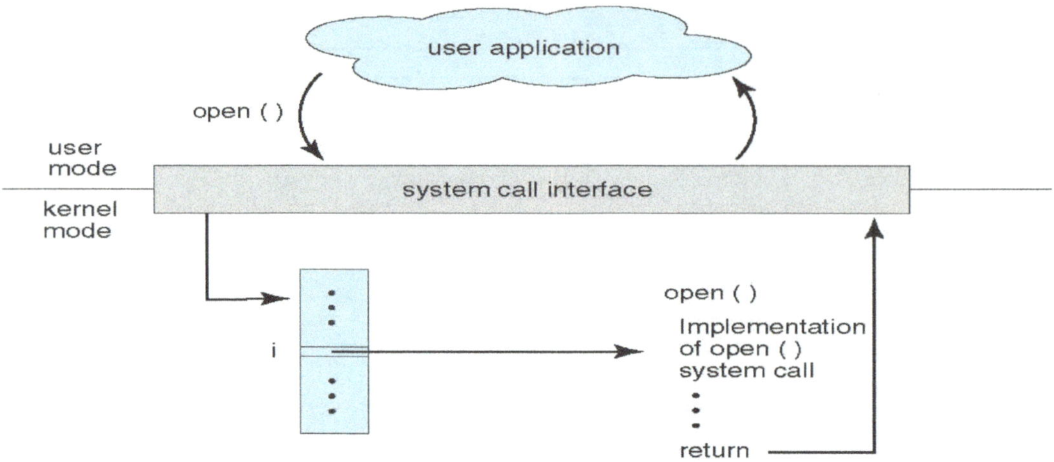

Figure1.36 The handling of a user application invoking the open()system call.

- Often, more information is required than simply identity of desired system call
  - Exact type and amount of information vary according to OS and call
- Three general methods used to pass parameters to the OS
  a. Simplest: pass the parameters in registers
     - In some cases, may be more parameters than registers
  b. Parameters stored in a block, or table, in memory, and address of block passed as a parameter in a register
     - This approach taken by Linux and Solaris
  c. Parameters placed, or pushed, onto the stack by the program and popped off the stack by the operating system
  d. Block and stack methods do not limit the number or length of parameters being passed

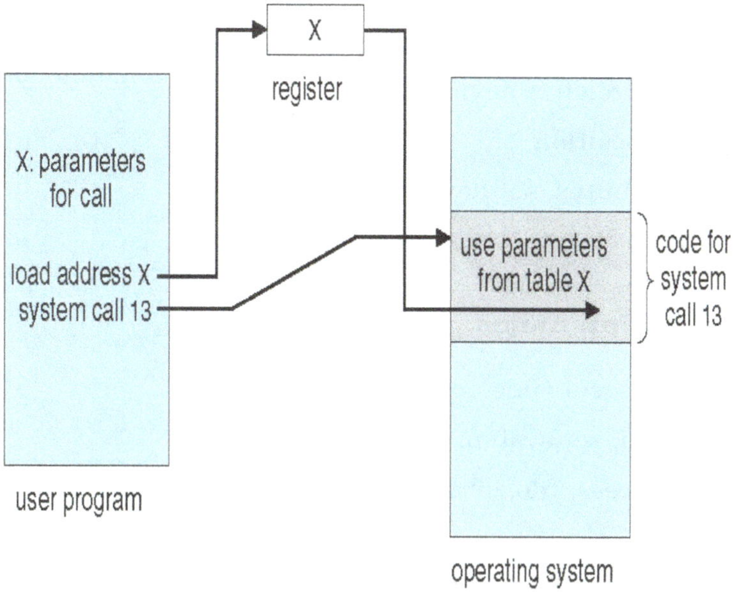

Figure 1.37 Passing parameters as a table

Types of System calls:

- **Process control**

  1. create process, terminate process
  2. end, abort
  3. load, execute
  4. get process attributes, set process attributes
  5. wait for time
  6. wait event, signal event
  7. allocate and free memory
  8. Dump memory if error
  9. Debugger for determining bugs, single step execution
  10. Locks for managing access to shared data between processes

- **File management**

  1. create file, delete file
  2. open, close file
  3. read, write, reposition
  4. get and set file attributes

- **Device management**
    1. request device, release device
    2. read, write, reposition
    3. get device attributes, set device attributes
    4. logically attach or detach devices
- **Information maintenance**
    1. get time or date, set time or date
    2. get system data, set system data
    3. get and set process, file, or device attributes
- **Communications**
    1. create, delete communication connection
    2. send, receive messages if message passing model to host name or process name
    3. From client to server
    4. Shared-memory model create and gain access to memory regions
    5. transfer status information
    6. attach and detach remote devices.
- **Protection**
    1. Control access to resources
    2. Get and set permissions
    3. Allow and deny user access

Examples of Windows and Unix System Calls

	Windows	Unix
Process Control	CreateProcess() ExitProcess() WaitForSingleObject()	fork() exit() wait()
File Manipulation	CreateFile() ReadFile() WriteFile() CloseHandle()	open() read() write() close()
Device Manipulation	SetConsoleMode() ReadConsole() WriteConsole()	ioctl() read() write()
Information Maintenance	GetCurrentProcessID() SetTimer() Sleep()	getpid() alarm() sleep()
Communication	CreatePipe() CreateFileMapping() MapViewOfFile()	pipe() shmget() mmap()
Protection	SetFileSecurity() InitlializeSecurityDescriptor() SetSecurityDescriptorGroup()	chmod() umask() chown()

1.13 SYSTEM PROGRAMS

- System programs provide a convenient environment for program development and execution. They can be divided into:
    i. File manipulation
    ii. Status information sometimes stored in a File modification
    iii. Programming language support
    iv. Program loading and execution
    v. Communications
    vi. Background services
    vii. Application programs
- Most users' view of the operation system is defined by system programs, not the actual system calls
- Provide a convenient environment for program development and execution
- Some of them are simply user interfaces to system calls; others are considerably more complex

1. **File management** - Create, delete, copy, rename, print, dump, list, and generally manipulate files and directories
2. **Status information**

a. Some ask the system for info - date, time, amount of available memory, disk space, number of users
b. Others provide detailed performance, logging, and debugging information
c. Typically, these programs format and print the output to the terminal or other output devices
d. Some systems implement a registry - used to store and retrieve configuration information

3. **File modification**
   a. Text editors to create and modify files
   b. Special commands to search contents of files or perform transformations of the text.
4. **Programming-language support** - Compilers, assemblers, debuggers and interpreters sometimes provided
5. **Program loading and execution**- Absolute loaders, relocatable loaders, linkage editors, and overlay-loaders, debugging systems for higher-level and machine language
6. **Communications** - Provide the mechanism for creating virtual connections among processes, users, and computer systems.

Allow users to send messages to one another's screens, browse web pages, send electronic-mail messages, log in remotely, transfer files from one machine to another

7. **Background Services**

   Launch at boot time

   - Some for system startup, then terminate
   - Some from system boot to shutdown
     a. Provide facilities like disk checking, process scheduling, error logging, printing
     b. Run in user context not kernel context
     c. Known as services, subsystems, daemons

8. **Application programs**
   a. Don't pertain to system
   b. Run by users
   c. Not typically considered part of OS
   d. Launched by command line, mouse click, finger poke.

1.15 OPERATING SYSTEM GENERATION AND SYSTEM BOOT

OPERATING SYSTEM GENERATION

Operating systems are designed to run on any of a class of machines; the system must be configured for each specific computer site

System generation SYSGEN program obtains information concerning the specific configuration of the hardware system

- Used to build system-specific compiled kernel or system-tuned
- Can general more efficient code than one general kernel.

The following kinds of information must be determined:

1. What CPU is to be used? What options (extended instruction sets, floating point arithmetic, and so on) are installed? For multiple CPU systems, each CPU may be described.
2. How will the boot disk be formatted? How many sections, or "partitions," will it be separated into, and what will go into each partition? How much memory is available? Some systems will determine this value themselves by referencing memory location after memory location until an "illegal address" fault is generated.
3. What devices are available? The system will need to know how to address each device (the device number), the device interrupt number, the device's type and model, and any special device characteristics.
4. What operating-system options are desired, or what parameter values are to be used? These options or values might include how many buffers of which sizes should be used, what type of CPU-scheduling algorithm are desired, what the maximum number of processes to be supported is, and so on.

Once this information is determined, it can be used in several ways.

At one extreme, a system administrator can use it to modify a copy of the source code of the operating system. The operating system then is completely compiled.

Selection allows the library to contain the device drivers for all supported I/O devices, but only those needed are linked into the operating system. Because the system is not recompiled, system generation is faster, but the resulting system may be overly general.

At the other extreme, it is possible to construct a system that is completely table driven. All the code is always part of the system, and selection occurs at execution time, rather than at compile or link time. System generation involves simply creating the appropriate tables to describe the system.

The major differences among these approaches are the size and generality of the generated system and the ease of modifying it as the hardware configuration changes.

SYSTEM BOOT

After an operating system is generated, it must be made available for use by the hardware. But how does the hardware know where the kernel is or how to load that kernel? The procedure of starting a computer by loading the kernel is known as booting the system.

On most computer systems, a small piece of code known as the bootstrap program or bootstrap loader locates the kernel, loads it into main memory, and starts its execution. Some computer systems, such as PCs, use a two-step process in which a simple bootstrap loader fetches a more complex boot program from disk, which in turn loads the kernel.

When a CPU receives a reset event—for instance, when it is powered up or rebooted—the instruction register is loaded with a predefined memory location, and execution starts there. At that location is the initial bootstrap program. This program is in the form of read-only memory (ROM), because the RAM is in an unknown state at system startup.

ROM is convenient because it needs no initialization and cannot easily be infected by a computer virus. The bootstrap program can perform a variety of tasks.

- One task is to run diagnostics to determine the state of the machine. If the diagnostics pass, the program can continue with the booting steps.
- It can also initialize all aspects of the system, from CPU registers to device controllers and the contents of main memory. Sooner or later, it starts the operating system.

Some systems—such as cellular phones, tablets, and game consoles—store the entire operating system in ROM. Storing the operating system in ROM is suitable for small operating systems, simple supporting hardware, and rugged operation.

A problem with this approach is that changing the bootstrap code requires changing the ROM hardware chips. Some systems resolve this problem by using erasable programmable read-only memory (EPROM), which is read only except when explicitly given a command to become writable.

All forms of ROM are also known as firmware, since their characteristics fall somewhere between those of hardware and those of software.

Problem with firmware:

- Executing code there is slower than executing code in RAM.
- Some systems store the operating system in firmware and copy it to RAM for fast execution.
- It is relatively expensive, so usually only small amounts are available.
- For large operating systems (including most general-purpose operating systems like Windows, Mac OS X, and UNIX) or for systems that change frequently, the bootstrap loader is stored in firmware, and the operating system is on disk.

In this case, the bootstrap runs diagnostics and has a bit of code that can read a single block at a fixed location (say block zero) from disk into memory and execute the code from that boot block.

The program stored in the boot block may be sophisticated enough to load the entire operating system into memory and begin its execution.

More typically, it is simple code (as it fits in a single disk block) and knows only the address on disk and length of the remainder of the bootstrap program.

GRUB is an example of an open-source bootstrap program for Linux systems. All of the disk-bound bootstrap, and the operating system itself, can be easily changed by writing new versions to disk.

A disk that has a boot partition is called a boot disk or system disk.

Now that the full bootstrap program has been loaded, it can traverse the file system to find the operating system kernel, load it into memory, and start its execution. It is only at this point that the system is said to be running.

REVIEW QUESTIONS

PART-A

1. What is meant by Operating Systems?

An Operating System (OS) exploits the hardware resources of one or more processors to provide a set of services to system users. It manages secondary memory and Input/Output devices.

(or)

An Operating System is a program that controls the execution of application programs. It acts as an interface between applications and the computer hardware.

2. List the basic element of a computer.

A computer consists of four main structural elements includes

a. Processor

b. Main memory

c. I/O modules

d. System bus

These components are interconnected with each other to achieve the function of the computer to execute the programs.

3. What is meant by serial processing?

1. No operating system
   - Programmers interacted directly with the computer hardware
2. Computers ran from a console with display lights, toggle switches, some form of input device, and a printer
3. Users have access to the computer in "series".

4. List out the drawbacks of serial processing

1. Scheduling
2. Setup time

5. List out the evolutions of a microprocessor.

1. Microprocessors
2. Graphical Processing Unit (GPS)
3. Digital Signal Processing (DSPs)
4. System on Chip (SoC).

6. What are the steps to be followed for an instruction to be executed?

A program to be executed by a processor consists of a set of instructions stored in memory. It consists of two steps:

1. The processor reads (fetches) instructions from memory one at a time (Fetch Stage)
2. Executes each instruction (Execute Stage).

7. Draw the basic instruction cycle.

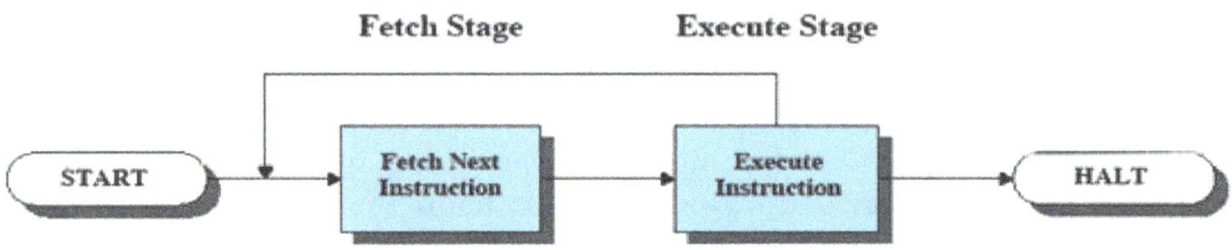

8. What are the four categories of actions according to instructions in instruction register(IR)?

- **Processor-memory**: Data may be transferred from processor to memory or from memory to processor.
- **Processor-I/O**: Data may be transferred to or from a peripheral device by transferring between the processor and an I/O module.
- **Data processing**: The processor may perform some arithmetic or logic operation on data.
- **Control:** An instruction may specify that the sequence of execution be altered.

9. List out the Internal CPU registers.

1. Program Counter (PC)
2. Accumulator (AC)
3. Instruction Register (IR).

10. What is meant by Interrupt? List the classes of interrupts.

Interrupts are provided primarily a way to improve processor utilization. The most classes of interrupts are:

1. Program
2. Timer
3. Input/Output
4. Hardware failure

11. What are the three sections of I/O program?

- **A sequence of instructions used** to prepare for the actual I/O operation.
- **The actual I/O command**. Without the use of interrupts, once this command is issued, the program must wait for the I/O device to perform the requested function
- **A sequence of instructions**, used to complete the operation, which includes setting a flag indicating the success or failure of the operation.

12. List the types of interrupt.

1. Single interrupts
2. Multiple interrupts.

13. What is meant by Program Status Word (PWS)?

The PSW contains status information about the currently running process, including memory usage information, condition codes, and other status information, such as an interrupt enable/disable bit and a kernel/user mode bit.

14. How multiple interrupts are handled in a system?

1. Sequential interrupt processing (disable interrupts while an interrupt is being processed)
2. Nested interrupt processing (use a priority scheme)

15. Drawbacks of sequential interrupt processing.

1. It does not take into account relative priority or time-critical needs.
2. If the first batch of input has not been processed before the second batch arrives, data may be lost because the buffer on the I/O device may fill and overflow.

16. Define Principle of Locality.

The basis of the validity of "decreasing frequency of access to the memory by the processor" is a principle known as locality of reference. This principle can be applied across more than two levels of memory.

17. Suppose that the processor has access to two levels of memory. Level 1 contains 1,000 bytes and has an access time of 0.1 µs; level 2 contains 100,000 bytes and has an access time of 1 µs. Calculate the average access time of the system if hit ratio is 95 %

$$(0.95)\ (0.1\ \mu s) + (0.05)\ (0.1\ \mu s + 1\ \mu s) = 0.095 + 0.055 = 0.15\ \mu s$$

18. Define Cache memory.

1. Invisible to the OS
2. Interacts with other memory management hardware
3. Processor must access memory at least once per instruction cycle
4. Processor execution is limited by memory cycle time
5. Exploit the principle of locality with a small, fast memory

19. What is the basic principle of cache?

1. The cache contains a copy of a portion of main memory.
2. When the processor attempts to read a byte or word of memory, a check if the byte or word is in the cache.
    - If so, the byte or word is delivered to the processor.
    - If not, a block of main memory, is read into the cache and then the byte or word is delivered to the processor.

20. List the cache design categories.

The cache design fall into the following categories:

1. Cache size
2. Block size
3. Mapping function
4. Replacement algorithm
5. Write policy
6. Number of cache levels

21. List out the levels of cache.

1. Finally, a common place to have multiple levels of cache
2. labeled L1 (cache closest to the processor)
3. L2, and in many cases a third level L3

22. What are the constraints affect the design of the mapping function in cache?

Two constraints affect the design of the mapping function.

1. When one block is read in, another may have to be replaced. Replace a block that will not be needed in the near future, and has to maximize the hit ratio.
2. The more flexible the mapping function, the more complex is the circuitry required to search the cache to determine if a given block is in the cache.

23. What is meant by write policy?

1. If the contents of a block in the cache are altered, it is necessary to write it back to main memory before replacing it.
2. The write policy dictates when the memory write operation takes place.
1. At one extreme, the writing can occur every time that the block is updated.
2. At the other extreme, the writing occurs only when the block is replaced.
3. The latter policy minimizes memory write operations but leaves main memory in an obsolete state. This can interfere with multiple-processor operation and with direct memory access by I/O hardware modules.

24. List out the possible techniques used for I/O operations.

1. Programmed I/O
2. Interrupt-Driven I/O
3. Direct Memory Access (DMA).

25. What is meant by programmed I/O?

In the case of programmed I/O,

1. The I/O module performs the requested action and then sets the appropriate bits in the I/O status register but takes no action to alert the processor. It does not interrupt the processor.
2. To know the status of completion, the processor periodically checks the status of the I/O module until it finds that the operation is complete.
3. Hence, the processor has to wait a long time for the I/O module to know the status and must repeatedly interrogate the status of the I/O module.
4. Hence, the performance level of the entire system is severely degraded.

27. What is meant by interrupt driven I/O?

Interrupt-driven I/O

1. It is an alternative, known as interrupt-driven I/O
2. It issues an I/O command to a module and then go on to do some other useful work.
3. The I/O module will then interrupt the processor to request service when it is ready to exchange data with the processor.
4. The processor then executes the data transfer, as before, and then resumes its former processing.
5. More efficient than the programmed I/O.

29. List out the drawbacks of programmed I/O and interrupt driven I/O.

1. The I/O transfer rate is limited by the speed with which the processor can test and service a device.
2. The processor is tied up in managing an I/O transfer; a number of instructions must be executed for each I/O transfer.

30. List the purpose of using Direct Memory Access (DMA).

When large volumes of data are to be moved an efficient technique "Direct Memory Access" is used. The data can be moved by DMA either by

- Separate module on the system bus (or)
- Can be incorporated into an I/O module

31. Give the approaches supports for parallel execution.

1. Symmetric multiprocessors (SMPs)
2. Multicore computers
3. Clusters.

32. List out the advantages of SMP.

1. Performance
2. Availability
3. Incremental growth
4. Scaling

33. What is meant by Multicore processors?

1. A Multicore computer, also known as a chip multiprocessor , combines two or more processors (called cores) on a single piece of silicon (called a die).
2. Each core consists of all of the components of an independent processor, such as registers, ALU, pipeline hardware, and control unit, plus L1 instruction and data caches.
3. Also includes L2 cache and, in some cases, L3 cache.

34. What is the main objective of an operating system?

1. **Convenience**: An OS makes a computer more convenient to use.
2. **Efficiency**: An OS allows the computer system resources to be used in an efficient manner.
3. **Ability to evolve**: An OS should be constructed in such a way as to permit the effective development, testing, and introduction of new system functions without interfering with service.

35. List the services of Operating Systems.

1. Program development
2. Program Execution
3. Access to I/O devices
4. Controlled access to files
5. System access
6. Error detection and response
7. Accounting

36. List out the key interfaces of OS

1. Instruction set architecture (ISA)
2. Application binary interface (ABI)
3. Application programming interface (API)

37. What is a monitor?

1. The central idea behind the simple batch-processing scheme is the use of a piece of software known as the monitor
2. Monitor controls the sequence of events
3. Resident Monitor is software always in memory (available for execution)
4. Monitor reads in job and gives control
5. When job completes, Job returns control to monitor

38. What is meant by JCL?

Job Control Language (JCL)

1. Special type of programming language used to provide instructions to the monitor to improve job set up time
2. what compiler to use
3. what data to use

39. What are the overhead of simple batch system?

1. Processor time alternates between execution of user programs and execution of the monitor

2. **Sacrifices:**
   - some main memory is now given over to the monitor
   - some processor time is consumed by the monitor
3. Most of the time the processor is in idle state.

40. Differentiate batch multiprogramming and time sharing systems.

	Batch Multiprogramming	Time Sharing
Principle Objective	Maximize processor use	Minimize processor use
Source of directives to operating system	Job control language commands provided with the job	Commands entered at the terminal

41. Define time slicing.

A system clock generated interrupts at a rate of approximately one every 0.2 seconds. At each clock interrupt, the OS regained control and could assign the processor to another user. This technique is known as time slicing.

42. Draw the four components of a computer system.

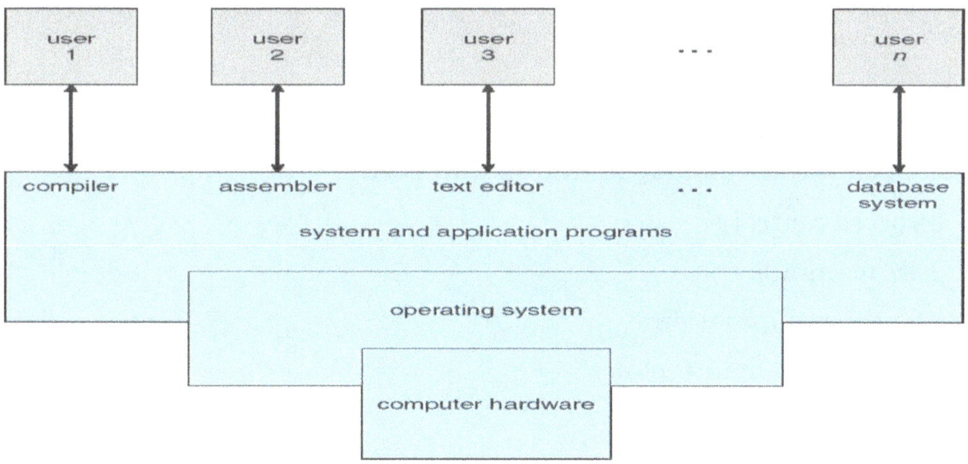

43. Define job pool.

Main memory is too small to accommodate all jobs. Hence the jobs are kept initially on the disk in the job pool. The job pool consists of all processes residing on disk awaiting allocation of main memory. The OS picks and begins to execute one of the jobs in memory.

44. What is meant by system call?

System calls provide an interface to the services made available by an operating system.

1. Programming interface to the services provided by the OS
2. Typically written in a high-level language (C or C++)
3. Mostly accessed by programs via a high-level Application Programming Interface (API) rather than direct system call use

45. How are the parameters passed in system calls?

- Three general methods used to pass parameters to the OS
  1. Simplest: pass the parameters in registers
     - In some cases, may be more parameters than registers
  2. Parameters stored in a block, or table, in memory, and address of block passed as a parameter in a register
     - This approach taken by Linux and Solaris
  3. Parameters placed, or pushed, onto the stack by the program and popped off the stack by the operating system
     - Block and stack methods do not limit the number or length of parameters being passed

46. List the types of system call.

System calls can be grouped into six categories:

1. Process control
2. File manipulation
3. Device manipulation
4. Information maintenance
5. Communications
6. Protection

47. What is meant by system program or system utilities?

System programs, also known as system utilities, provide a convenient environment for program development and execution. It can be divided into the following categories:

1. File management
2. Status information
3. File modification
4. Programming-language support
5. Program loading and execution
6. Communication
7. Background services

48. What is the purpose of bootstrap program or bootstrap loader?

On most computer systems, a small piece of code known as the bootstrap program or bootstrap loader locates the kernel, loads it into main memory and starts its execution.

PART-B

1. Explain in detail about the instruction execution with suitable example.
2. Explain the following:
   i. Program execution - no interrupt with suitable example
   ii. Simple interrupt-Short I/O wait with suitable example
   iii. Simple interrupt-Long I/O wait with suitable example
   iv. Interrupt Processing steps
   v. Multiple interrupts
3. Explain about the memory hierarchy in detail.
4. Explain about the cache principle, structure of cache/main memory and its design
5. categories in detail.
6. Write short notes about the Direct Memory Access (DMA).
   i. Write short notes on
   ii. Symmetric Multiprocessors (SMPs)-
7. Multicore Computers
   i. Write short notes on
   ii. Operating system as a user/computer interface
8. Operating System as a resource manager
9. Discuss in detail about the evolution of operating system.

10. Explain in detail about the computer system organization.
11. Explain about the operating system structure and operations in detail.
12. Write short notes on system calls.
13. Write short notes on -System programs, System boot

2

PROCESS MANAGEMENT

2.1 PROCESES AND PROCESS CONCEPT

1. An operating system executes a variety of programs:
   - Batch system – jobs
2. Time-shared systems – user programs or tasks
3. Textbook uses the terms job and process almost interchangeably.

Process – a program in execution; process execution must progress in sequential fashion

It has Multiple parts

1. The program code, also called text section
2. Current activity including program counter, processor registers
3. Stack containing temporary data
4. Function parameters, return addresses, local variables
5. Data section containing global variables
6. Heap containing memory dynamically allocated during run time
   - Program is passive entity stored on disk (executable file), process is active
   - Program becomes process when executable file loaded into memory
7. Execution of program started via GUI mouse clicks, command line entry of its name, etc
   - One program can be several processes
   - Consider multiple users executing the same program

Process in memory

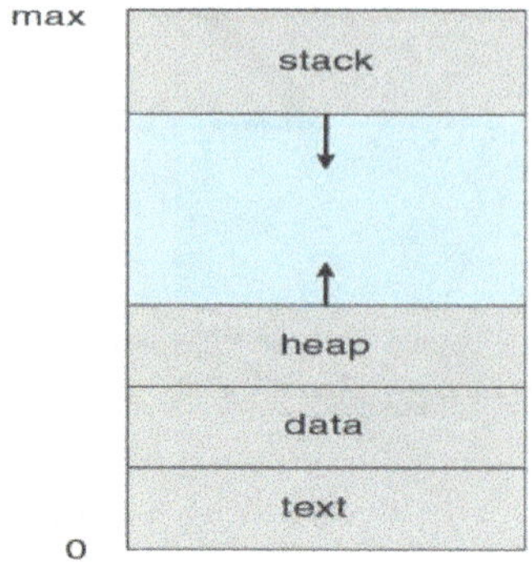

Figure 2.1 Process in memory.

2.1.1 Process States

As a process executes, it changes state. The state of a process is defined in part by the current activity of that process. A process may be in one of the following states:

1. **new**: The process is being created
2. **running**: Instructions are being executed
3. **waiting**: The process is waiting for some event to occur
4. **ready:** The process is waiting to be assigned to a processor
5. **terminated**: The process has finished execution

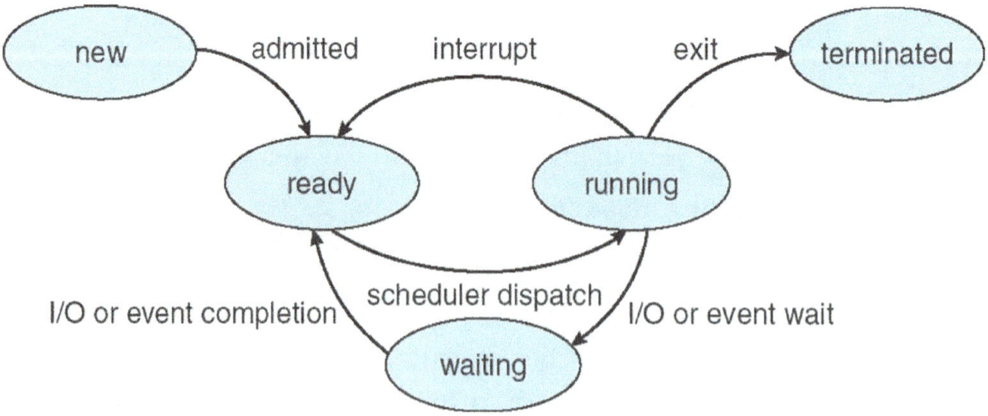

Figure 2.2 Process states

2.1.2 Process Control Block (PCB)

Each process is represented in the operating system by a process control block (PCB)—also called a task control block. A PCB is shown in Figure.

1. **Process state** –may be new, ready, running, waiting, terminated etc
2. **Program counter** – location of instruction to next execute
3. **CPU registers** – contents of all process-centric registers
4. **CPU scheduling information**- priorities, scheduling queue pointers
5. **Memory-management information** – memory allocated to the process. This information may include such items as the value of the base and limit registers and the page tables, or the segment tables, depending on the memory system used by the operating system
6. **Accounting information** – the amount of CPU and real time used, time limits, account numbers, job or process numbers, and so on.
7. **I/O status information** – I/O devices allocated to process, list of open files and so on.

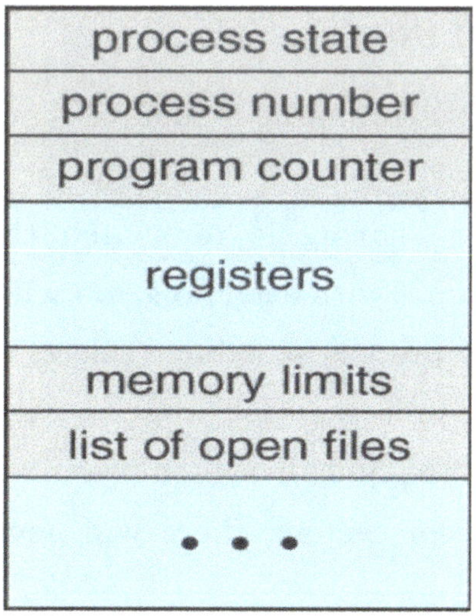

Figure 2.3 Process control block (PCB).

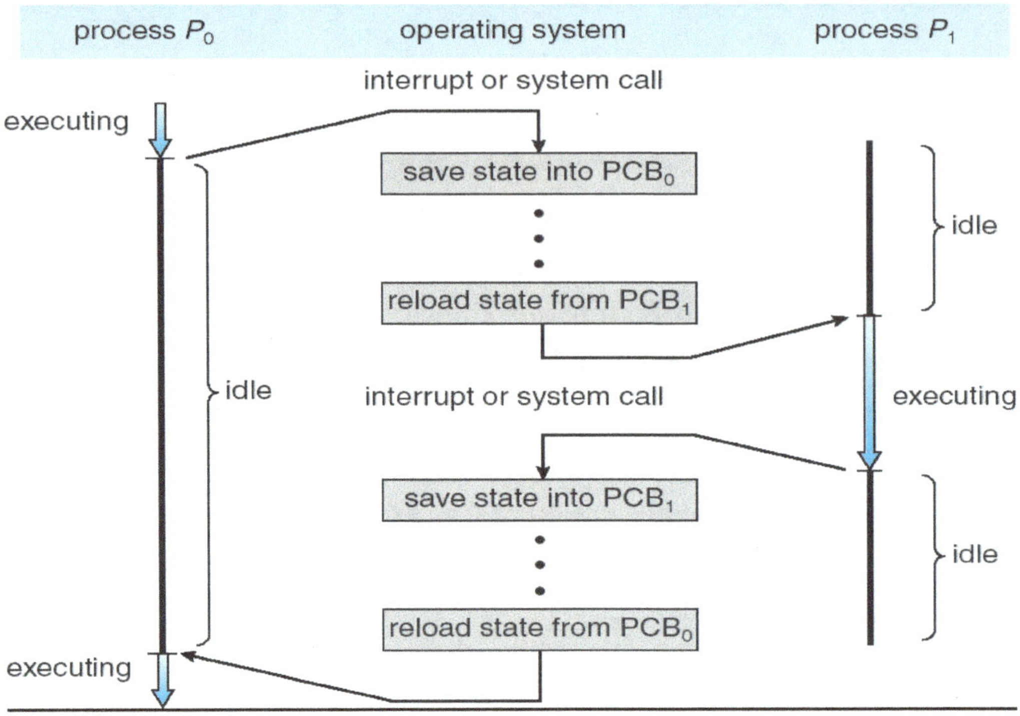

Figure 2.4 Diagram showing CPU switch from process to process

2.2 PROCESS SCHEDULING

The objective of multiprogramming is to have some process running at all times, to maximize CPU utilization.

The objective of time sharing is to switch the CPU among processes so frequently that users can interact with each program while it is running.
To meet these objectives, the process scheduler selects an available process (possibly from a set of several available processes) for program execution on the CPU.

- To Maximize CPU use, quickly switch processes onto CPU for time sharing
- For a single-processor system, there will never be more than one running process.
- If there are more processes, the rest have to wait until the CPU is free and can be rescheduled.

2.2.1 Scheduling Queues

- Maintains scheduling queues of processes
  1. **Job queue** – set of all processes in the system
  2. **Ready queue** – set of all processes residing in main memory, ready and waiting to execute
  3. **Device queues** – set of processes waiting for an I/O device Processes migrate among the various queues

Ready Queue and Various I/O Device Queues

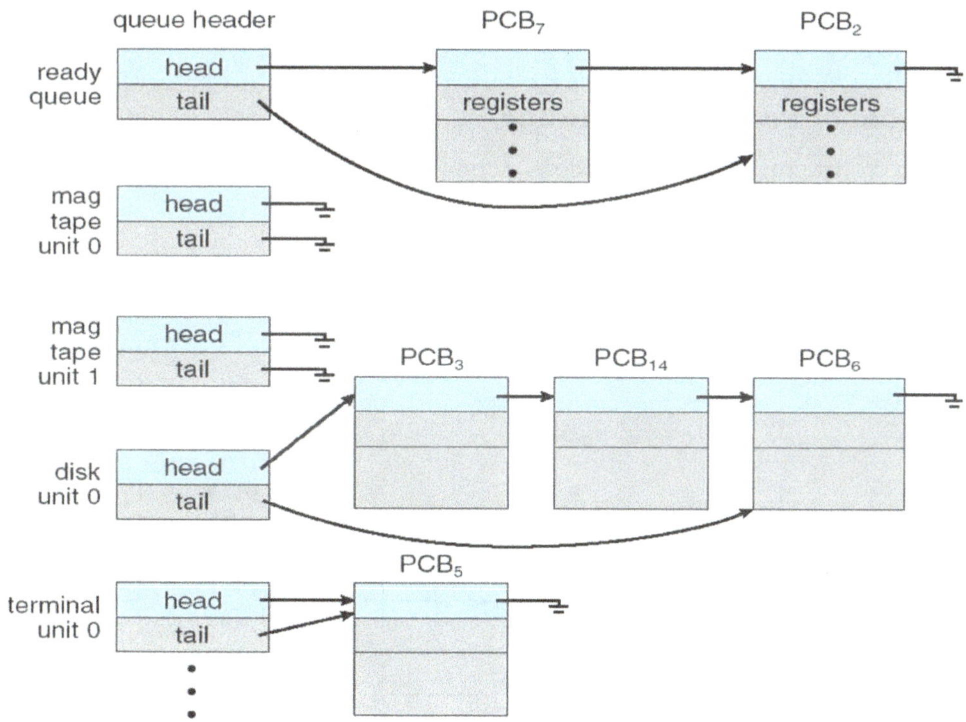

Figure 2.5 The ready queue and various I/O device queues.

As processes enter the system, they are put into a job queue, which consists of all processes in the system.

The processes that are residing in main memory and are ready and waiting to execute are kept on a list called the ready queue. This queue is generally stored as a linked list.

A ready-queue header contains pointers to the first and final PCBs in the list. Each PCB includes a pointer field that points to the next PCB in the ready queue.

The system also includes other queues. When a process is allocated the CPU, it executes for a while and eventually quits, is interrupted, or waits for the occurrence of a particular event, such as the completion of an I/O request.

Suppose the process makes an I/O request to a shared device, such as a disk. Since there are many processes in the system, the disk may be busy with the I/O request of some other process.

The process therefore may have to wait for the disk. The list of processes waiting for a particular I/O device is called a device queue. Each device has its own device queue.

Representation of Process Scheduling

A common representation of process scheduling is a queueing diagram. Queueing diagram represents queues, resources, flows.

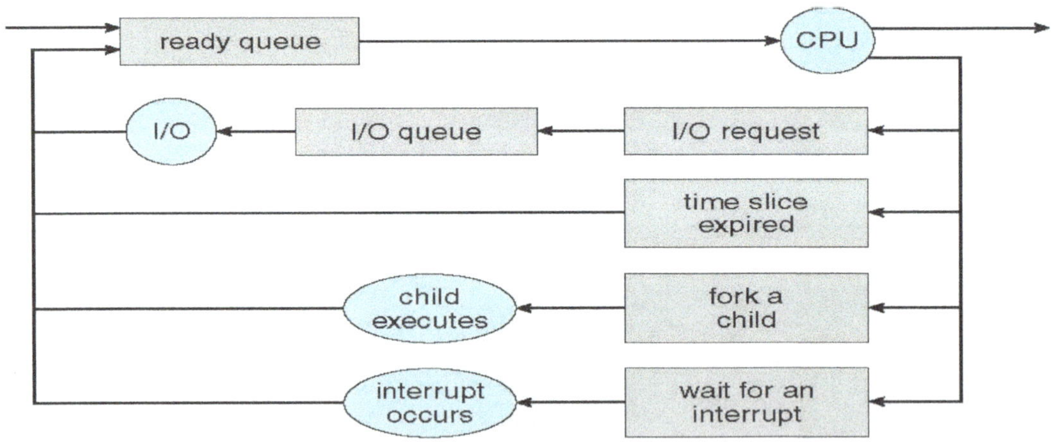

Figure 2.6 Queueing-diagram representation of process scheduling.

Two types of queues are present:

- Ready queue and a set of device queues.
- The circles represent the resources that serve the queues, and the arrows indicate the flow of processes in the system.
- A new process is initially put in the ready queue. It waits there until it is selected for execution, or dispatched.

- Once the process is allocated the CPU and is executing, one of several events could occur:
- The process could issue an I/O request and then be placed in an I/O queue.
- The process could create a new child process and wait for the child's termination.
- The process could be removed by force from the CPU, as a result of an interrupt, and be put back in the ready queue.

The process switches from the waiting state to the ready state and is then put back in the ready queue. This cycle until it terminates, at which time it is removed from all queues and has its PCB and resources deallocated.

2.2.2 Schedulers

The operating system selects processes from queues for scheduling. The selection process is carried out by the appropriate scheduler.

Two types of schedulers:

1. **Short Term Schedulers (STS)** or **CPU Scheduler**
2. **Long Term Schedulers** (LTS) or **Job Scheduler**

1. **Short-term scheduler** (or) **CPU scheduler** – selects which process should be executed next and allocates CPU
    - Sometimes the only scheduler in a system
    - Short-term scheduler is invoked frequently (milliseconds) ⇒ (must fast)
    - Selects from among the processes that are ready to execute and
    - allocates the CPU to one of them.
    - A process may execute for only a few milliseconds before waiting for an I/O request.
    - Executes at least once every 100 milliseconds.
    - Because of the short time between executions, it must be fast. For Ex: If it takes 10 milliseconds to decide to execute a process for 100 milliseconds, then 10/(100 + 10) = 9 percent of the CPU is being used (wasted) simply for scheduling the work.

2. **Long-term scheduler** (or) **job scheduler** – selects which processes should be brought into the ready queue
    - Long-term scheduler is invoked infrequently (seconds, minutes) ⇒ (may be slow)
    - The long-term scheduler controls the degree of multiprogramming

In general, most processes can be described as
- Either I/O bound or
- CPU bound.
- I/O-bound process – spends more time doing I/O than computations, many short CPU bursts
- CPU-bound process – spends more time doing computations; few very long CPU bursts

Long-term scheduler strives for good process mix

Addition of Medium Term Scheduling

The long-term scheduler selects a good process mix of I/O-bound and CPU-bound processes.

- If all processes are I/O bound - the ready queue will almost always be empty, and the short-term scheduler will have little to do.
- If all processes are CPU bound - the I/O waiting queue will almost always be empty, devices will go unused, and again the system will be unbalanced.

Some operating systems, such as time-sharing systems, may introduce an additional, intermediate level of scheduling called medium-term scheduler.

The key idea behind a medium-term scheduler - it can be advantageous to remove a process from memory (and from active contention for the CPU) and thus reduce the degree of multiprogramming.

Later, the process can be reintroduced into memory, and its execution can be continued where it left off. This scheme is called swapping.

The process is swapped out, and is later swapped in, by the medium-term scheduler. Swapping may be necessary to improve the process mix.

Medium-term scheduler can be added if degree of multiple programming needs to decrease

- Remove process from memory, store on disk, bring back in from disk to continue execution: swapping

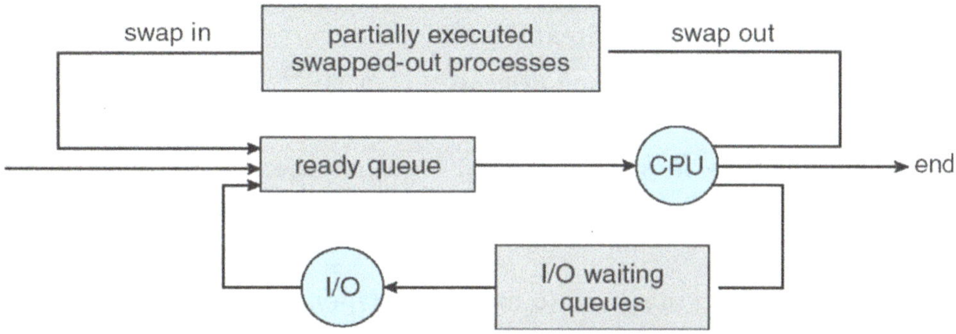

Figure 2.7 Addition of medium-term scheduling to the queueing diagram.

Multitasking in Mobile Systems

- Some mobile systems (e.g., early version of iOS) allow only one process to run, others suspended
- Due to screen real estate, user interface limits iOS provides for a
  - ➢ Single foreground process- controlled via user interface
  - ➢ Multiple background processes– in memory, running, but not on the display, and with limits
  - ➢ Limits include single, short task, receiving notification of events, specific long-running tasks like audio playback
- Android runs foreground and background, with fewer limits
  - ➢ Background process uses a service to perform tasks
  - ➢ Service can keep running even if background process is suspended
  - ➢ Service has no user interface, small memory use

2.2.3 Context Switch

- Switching the CPU to another process requires performing a state save of the current process and a state restore of a different process. This task is known as a context switch i.e., When CPU switches to another process, the system must save the state of the old process and load the saved state for the new process via a context switch
- Context of a process represented in the PCB

Context-switch time is pure overhead, because the system does no useful work while switching.

Switching speed varies from machine to machine, depending on

- The memory speed,
- The number of registers that must be copied, and
- The existence of special instructions (such as a single instruction to load or store all registers).

The more complex the OS and the PCB, the longer the context switch.

2.3 OPERATIONS ON PROCESSES

- The processes in most systems can execute concurrently, and they may be created and deleted dynamically. Thus, system must provide mechanisms for:
  - Process creation
  - Process termination

2.3.1 Process Creation

- Generally, process identified and managed via a process identifier (pid)
- During the course of execution, a process may create several new processes.
- The creating process is called a parent process, and
- The new processes are called the children of that process.
- Each of these new processes may in turn create other processes, forming a tree of processes.

- Most operating systems (including UNIX, Linux, and Windows) identify processes according to a unique process identifier (or pid), which is typically an integer number.

Figure illustrates a typical process tree for the Linux operating system, showing the name of each process and its pid. The init process (which always has a pid of 1) serves as the root parent process for all user processes.

Once the system has booted, the init process can also create various user processes, such as a web or print server, an ssh server, and the like. There are two children of init—k threadd and sshd.

- The kthreadd process is responsible for creating additional processes that perform tasks on behalf of the kernel (in this situation, khelper and pdflush).
- The sshd process is responsible for managing clients that connect to the system by using ssh(which is short for secure shell).

The login process is responsible for managing clients that directly log onto the system.

In this example, a client has logged on and is using the bash shell, which has been assigned pid 8416.

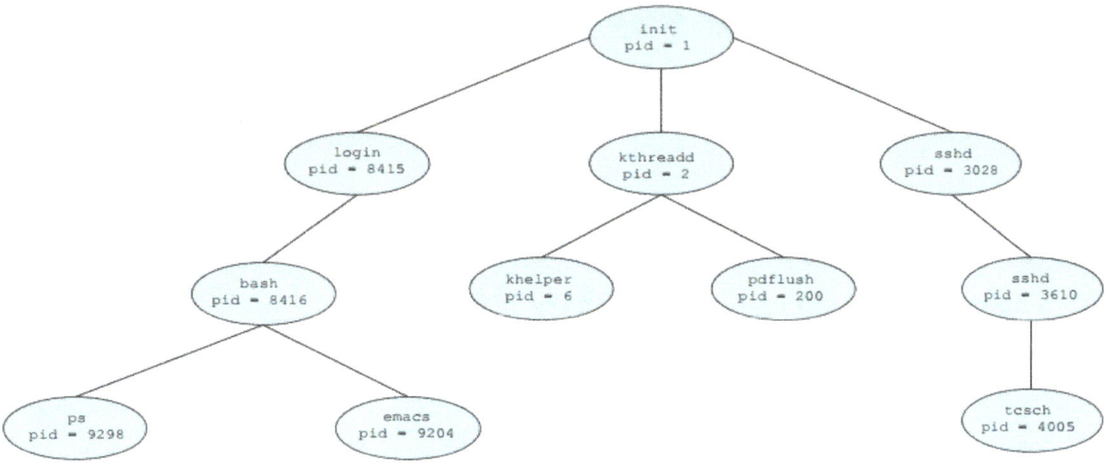

Figure 2.8 A tree of processes on a typical linux system.

- Resource sharing options
  - ➢ Parent and children share all resources
  - ➢ Children share subset of parent's resources
  - ➢ Parent and child share no resources

When a process creates a new process, two possibilities for execution exist:

1. The parent continues to execute concurrently with its children.
2. The parent waits until some or all of its children have terminated.

There are also two address-space possibilities for the new process:

1. The child process is a duplicate of the parent process.
2. The child process has a new program loaded into it.

UNIX Examples

A new process is created by the fork () system call, consists of a copy of the address space of the original process.

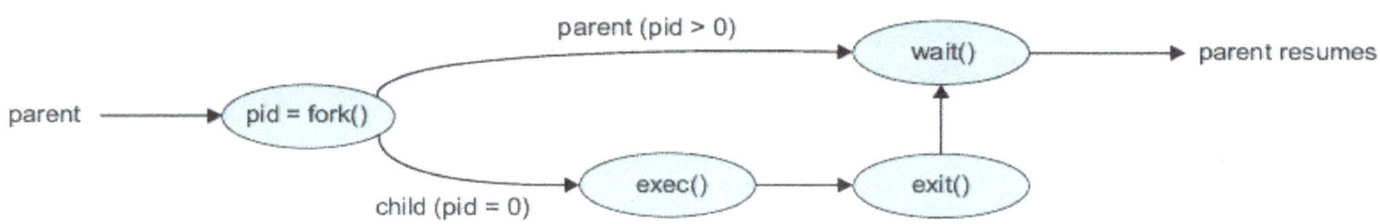

Figure 2.9 Process creation using the fork () system call

After a fork () system call, one of the two processes typically uses the exec () system call to replace the process's memory space with a new program.

The exec () system call loads a binary file into memory and starts its execution. In this manner, the two processes are able to communicate and then go their separate ways.

The parent can then create more children; or, if it has nothing else to do while the child runs, it can issue a wait () system call to move itself off the ready queue until the termination of the child.

The only difference is that the value of pid (the process identifier) for the child process is zero, while that for the parent is an integer value greater than zero.

Process Termination

A process terminates when it finishes executing its final statement and asks the operating system to delete it by using the exit () system call.

- At that point, the process may return a status value (typically an integer) to its parent process (via the wait () system call).
- All the resources of the process—including physical and virtual memory, open files, and I/O buffers—are deallocated by the operating system.

A parent may terminate the execution of one of its children for a variety of reasons, such as these:

- The child has exceeded its usage of some of the resources that it has been allocated.
- The task assigned to the child is no longer required.
- The parent is exiting, and the operating system does not allow a child to continue if its parent terminates.

If a process terminates (either normally or abnormally), then all its children must also be terminated. This phenomenon, referred to as cascading termination, is normally initiated by the operating system.

When a process terminates, its resources are deallocated by the operating system. However, its entry in the process table must remain there until the parent calls wait (), because the process table contains the process's exit status.

A process that has terminated, but whose parent has not yet called wait (), is known as a zombie process. All processes transition to this state when they terminate, but generally they exist as zombies. Once the parent calls wait (), the process identifier of the zombie process and its entry in the process table are released.

If a parent did not invoke wait () and instead terminated, thereby leaving its child processes as orphans.

2.4 INTER-PROCESS COMMUNICATION

- Processes executing concurrently in the operating system may be either independent processes or cooperating processes.
- A process is independent if it cannot affect or be affected by the other processes executing in the system. Any process that does not share data with any other process is independent.
- A process is cooperating if it can affect or be affected by the other processes executing in the system.
- Cooperating process can affect or be affected by other processes, including sharing data
- Reasons for cooperating processes:
    - **Information sharing** - an environment must be provided to allow concurrent access to information(shared file etc)
    - **Computation speedup** - To improve the speedup, tasks are divided into subtasks, and executes in parallel
    - **Modularity-** dividing the system functions into separate processes or threads
    - **Convenience-** Even an individual user may work on many tasks at the same time.
- Cooperating processes need inter-process communication (IPC)

Communications Models

- Two models of IPC
    - Shared memory
    - Message passing

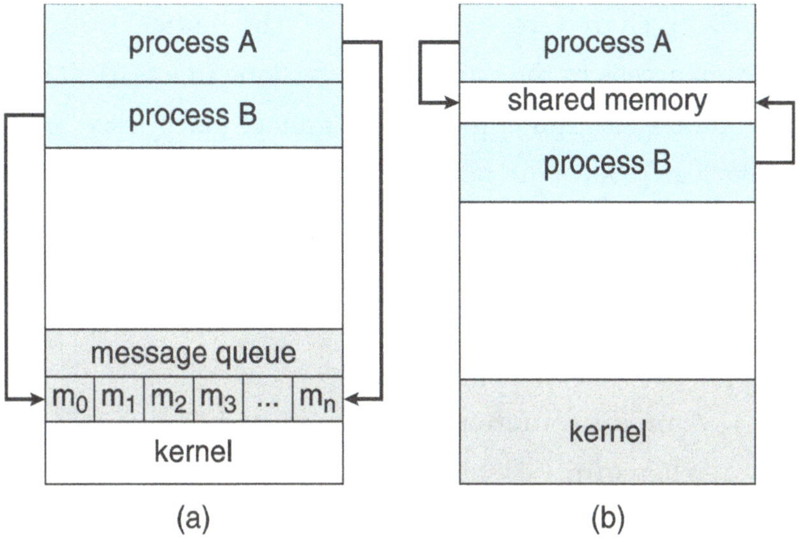

Figure 2.10 Communications models. (a) Message passing. (b) Shared memory.

2.4.1 Communication Model – Shared Memory

Interprocess communication using shared memory requires communicating processes to establish a region of shared memory that resides in the address space of the process creating the shared-memory segment. Other processes also attach it to their address space.

They can then exchange information by reading and writing data in the shared areas and are not under the operating system's control.

The processes are also responsible for ensuring that they are not writing to the same location simultaneously.

To illustrate the concept of cooperating processes, consider the producer–consumer problem, which is a common paradigm for cooperating processes.

- A producer process produces information that is consumed by a consumer process. For example, a compiler may produce assembly code that is consumed by an assembler. The assembler, in turn, may produce object modules that are consumed by the loader.
- Two types of processes, producers and consumers, wish to communicate.
- Producers have information to send to the consumers. They place the information in a bounded (limited size) buffer, where consumers can obtain it.

- The buffer is a shared resource; use of the buffer must be synchronized. Obviously each access to the buffer must be done in a critical section.
- When the buffer is full producers cannot place any more items in it. They therefore go to sleep, to be awakened when a consumer removes an item.
- Likewise, consumers go to sleep when the buffer is empty, to be awaked by a consumer placing an item in the buffer.

```
item nextproduced;
while (true){
/* produce an item in nextproduced */
while (((in + 1)% BUFFERSIZE) == out)
; /* do nothing */
buffer[in] = nextproduced;
in = (in + 1)% BUFFERSIZE;
}
```

The producer process using shared memory.
One solution to the producer–consumer problem uses shared memory.

To allow producer and consumer processes to run concurrently, a buffer of items must be available that can be filled by the producer and emptied by the consumer. This buffer will reside in a region of memory that is shared by the producer and consumer processes.

A producer can produce one item while the consumer is consuming another item. The producer and consumer must be synchronized, so that the consumer does not try to consume an item that has not yet been produced.

Two types of buffers can be used.

- The unbounded buffer places no practical limit on the size of the buffer. The consumer may have to wait for new items, but the producer can always produce new items.
- The bounded buffer assumes a fixed buffer size. In this case, the consumer must wait if the buffer is empty, and the producer must wait if the buffer is full.

To illustrate interprocess communication using shared memory, the following variables reside in a region of memory shared by the producer and consumer processes:

```
#define BUFFERSIZE 10
typedef struct {
...
}item;
item buffer[BUFFERSIZE];
int in = 0;
int out = 0;
```

The shared buffer is implemented as a circular array with two logical pointers: in and out.

The variable in points to the next free position in the buffer; out points to the first full position in the buffer.

The buffer is empty when in==out; the buffer is full when ((in+1) % BUFFERSIZE) ==out.

The solution is correct, but can only use BUFFER_SIZE-1 elements

The code for the producer process,

```
item next_produced;
while (true) {
        /* produce an item in next produced */
        while (((in + 1) % BUFFER_SIZE) == out)
                ; /* do nothing */
        buffer[in] = next_produced;
        in = (in + 1) % BUFFER_SIZE;
}
```

The code for the consumer process,

```
item nextconsumed;
while (true){
        while (in == out)
                ; /* do nothing */
        nextconsumed = buffer[out];
        out = (out + 1)% BUFFERSIZE;
        /* consume the item in nextconsumed */
}
```

The local variable nextproduced in which the new item to be produced is stored.

The consumer process has a local variable next consumed in which the item to be consumed is stored.

This scheme allows at most BUFFERSIZE−1 items in the buffer at the same time.

2.4.2 Communication Model – Message Passing

Message passing provides a mechanism to allow processes to communicate and to synchronize their actions without sharing the same address space.

It is particularly useful in a distributed environment, where the communicating processes may reside on different computers connected by a network. For example, an

Internet chat program could be designed so that chat participants communicate with one another by exchanging messages.

A message-passing facility provides at least two operations:

- send(message)
- receive(message)

Messages sent by a process can be either fixed or variable in size.

- If only fixed-sized messages can be sent, the system-level implementation is straightforward & makes the task of programming more difficult.
- Conversely, variable-sized messages require a more complex system- level implementation, but the programming task becomes simpler.

If processes P and Q want to communicate, they must send messages to and receive messages from each other: ie. a communication link must exist between them.

This link can be implemented in a variety of ways, for implementing a link and the send()/ receive() operations:

- Direct or indirect communication
- Synchronous or asynchronous communication
- Automatic or explicit buffering

Direct Communication

Each process that wants to communicate must explicitly name the recipient or sender of the communication.

In this scheme, the send () and receive () primitives are defined as:

- send (P, message)—Send a message to process P.
- receive (Q, message)—Receive a message from process Q.

A communication link in this scheme has the following properties:

- A link is established automatically between every pair of processes that want to communicate. The processes need to know only each other's identity to communicate.
- A link is associated with exactly two processes.
- Between each pair of processes, there exists exactly one link.
    1. This scheme exhibits symmetry in addressing; that is, both the sender process and the receiver process must name the other to communicate.
    2. A variant of this scheme employs asymmetry in addressing. Here, only the sender names the recipient; the recipient is not required to name the sender. In this scheme, the send() and receive() primitives are defined as follows:
        a. Send (P, message)—Send a message to process P.
        b. Receive (id, message)—Receive a message from any process. The variable id is set to the name of the process with which comunicn has taken place.

Disadvantage in both of these schemes (symmetric and asymmetric) is the limited modularity of the resulting process definitions.

All references to the old identifier must be found, so that they can be modified to the new identifier. In general, any such hard-coding techniques, where identifiers must be explicitly stated, are less desirable than techniques involving indirection.

Indirect communication

The messages are sent to and received from mailboxes, or ports. Each mailbox has a unique identification.

A process can communicate with another process via a number of different mailboxes, but two processes can communicate only if they have a shared mailbox.
The send () and receive () primitives are defined as follows:

- send(A, message)—Send a message to mailbox A.
- receive(A, message)—Receive a message from mailbox A.

A communication link has the following properties:

- A link is established between a pair of processes only if both members of the pair have a shared mailbox.
- A link may be associated with more than two processes.
- Between each pair of communicating processes, a number of different links may exist, with each link corresponding to one mailbox.
- Suppose that processes P1, P2, and P3 all share mailbox A.
- Process P1sends a message to A, while both P2 and P3 execute a receive () from A.
- Which process will receive the message sent byP1?

Solutions:

1. Allow a link to be associated with two processes at most.
2. Allow at most one process at a time to execute a receive () operation.
3. Allow the system to select arbitrarily which process will receive the message (that is, either P2 or P3, but not both, will receive the message).

The operating system then must provide a mechanism that allows a process to do the following:

1. Create a new mailbox.
2. Send and receive messages through the mailbox.
3. Delete a mailbox.

Synchronization

Message passing may be blocking or nonblocking— also known as synchronous and asynchronous.

Blocking is considered synchronous

- Blocking send. The sending process is blocked until the message is received
- Blocking receive. The receiver blocks until a message is available.

Non-blocking is considered asynchronous

- Nonblocking send. The sending process sends the message and resumes operation.
- Nonblocking receive. The receiver retrieves either a valid message or a null.

Different combinations of send () and receive () are possible. When both send () and receive () are blocking, a rendezvous is happened between the sender and the receiver.

The solution to the producer–consumer problem becomes trivial when used blocking send () and receive () statements.

The producer merely invokes the blockingsend () call and waits until the message is delivered to either the receiver or the mailbox. Likewise, when the consumer invokes receive (), it blocks until a message is available.

```
message next_produced;
        while(true){
        /* produce an item in next produced */
  send(next_produced);
   }
message next_consumed;
while (true) {
```

```
receive(next_consumed);
    /* consume the item in next consumed */
}
```

Buffering

Whether communication is direct or indirect, messages exchanged by communicating processes reside in a temporary queue.

```
message nextproduced;
while (true){
/* produce an item in nextproduced */
send(nextproduced);
}
```

The producer process using message passing.

Basically, such queues can be implemented in three ways:

Zero capacity:

The queue has a maximum length of zero; the link cannot have any messages waiting in it. So the sender must block until the recipient receives the message (rendezvous).

Bounded capacity:

The queue has finite length n; thus, at most n messages can reside in it.

If the queue is not full when a new message is sent, the message is placed in the queue and the sender can continue execution without waiting.

If the link is full, the sender must block until space is available in the queue.

Unbounded capacity:

The queue's length is potentially infinite; thus, any number of messages can wait in it.

The sender never blocks.

2.5 THREADS- OVERVIEW

A thread is a basic unit of CPU utilization:

- It comprises a thread ID, a program counter, a register set, and a stack.
- It shares with other threads belonging to the same process its code section, data section, and other operating-system resources, such as open files and signals.

2.5.1 Motivation

- Most software applications that run on modern computers are multithreaded.
- A separate process with several threads runs within application.
- If a process has multiple threads of control, it can perform more than one task at a time. Figure 2.10 illustrates the difference between a traditional single-threaded process and a multithreaded process.

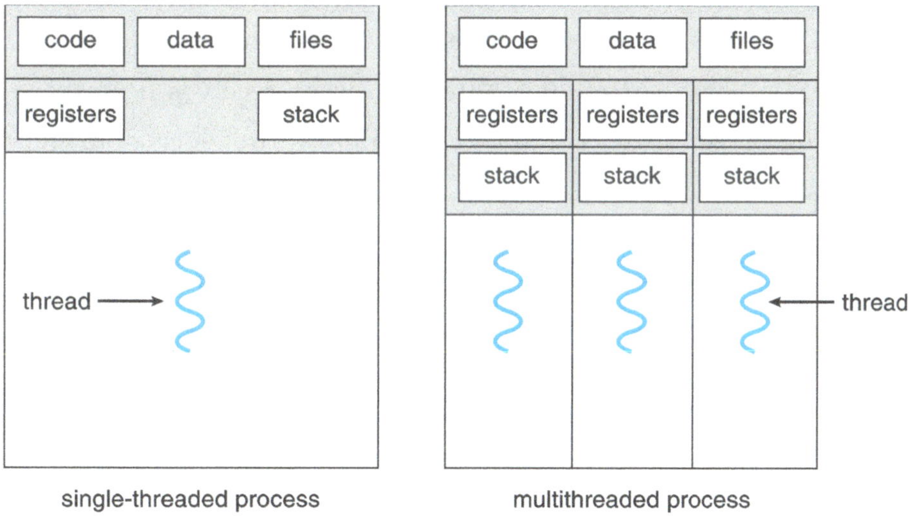

Figure 2.11 Single-threaded and multithreaded processes.

A web browser might have one thread display images or text while another thread retrieves data from the network,

For example,

- A word processor may have a thread for displaying graphics,
- Another thread for responding to keystrokes from the user,

- A third thread for performing spelling and grammar checking in the background.

When the server receives a request, it creates a separate process to service that request. This process-creation method was in common use before threads became popular.

Process creation is

- Time consuming and
- Resource intensive.

It is generally more efficient to use one process that contains multiple threads.

If the web-server process is multithreaded, the server will create a separate thread that listens for client requests. When a request is made, rather than creating another process, the server creates a new thread to service the request and resume listening for additional requests.

Threads also play a vital role in remote procedure call (RPC) systems.

RPC servers are multithreaded. When a server receives a message, it services the message using a separate thread. This allows the server to service several concurrent requests.

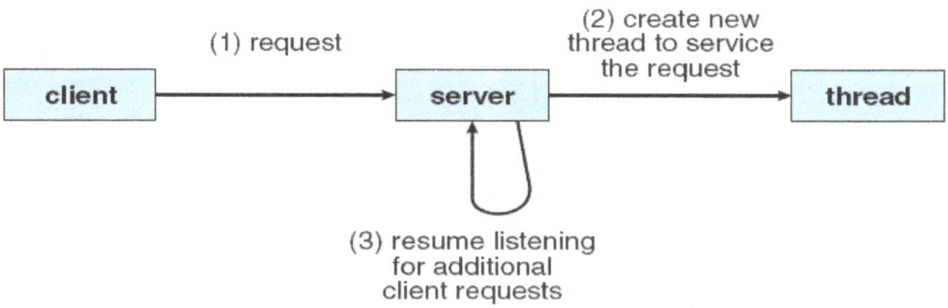

Figure 2.12 Multithreaded server architecture.

Finally, most operating-system kernels are multithreaded.

Several threads operate in the kernel, and each thread performs a specific task, such as managing devices, managing memory, or interrupt handling.

2.5.2 Benefits

1. **Responsiveness** – Multithreading an interactive application may allow continued execution if part of process is blocked, thereby increasing responsiveness to the user. Especially important for user interfaces.
2. **Resource Sharing** – threads share resources of process, easier than shared memory or message passing
3. **Economy** – cheaper than process creation, thread switching lower overhead than context switching
4. **Scalability** – process can take advantage of multiprocessor architectures, where threads may be running in parallel on different processing cores. A single-threaded process can run on only one processor, regardless how many are available.

2.6 MULTICORE PROGRAMMING

In Multicore or Multiprocessor Systems, each core appears as a separate processor to the operating system, whether the cores appear across CPU chips or within CPU chips. Multithreaded programming provides a mechanism for more efficient use of these multiple computing cores and improved concurrency.

- Parallelism implies a system can perform more than one task simultaneously
- Concurrency supports more than one task making progress.
    - ➢ Single processor / core, scheduler providing concurrency

Thus, it is possible to have concurrency without parallelism. CPU schedulers were designed to provide parallelism by rapidly switching between processes in the system, thereby allowing each process to make progress. Such processes were running concurrently, but not in parallel.

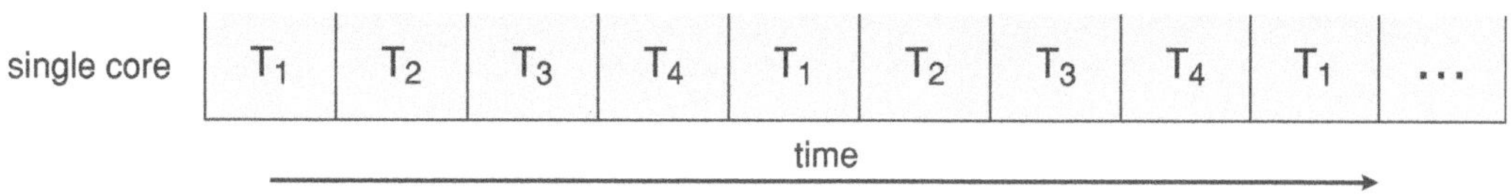

Figure 2.13 Concurrent execution on a single-core system

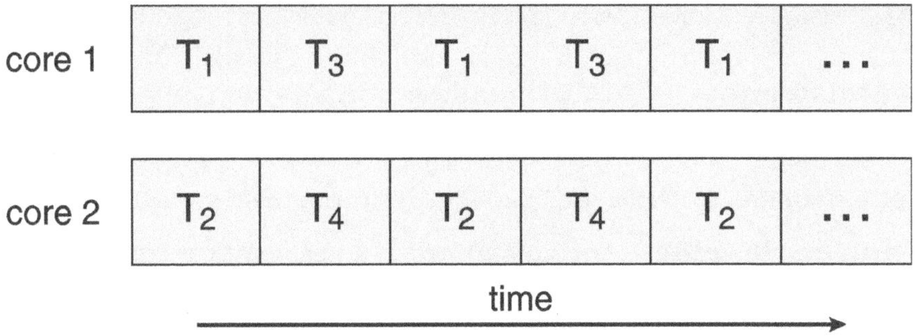

Figure 2.14 Parallel execution on a multicore system.

2.6.1 Programming Challenges

- Multicore or multiprocessor systems putting pressure on programmers, challenges include:
  - **Dividing activities**
    a. Divide the tasks into separate concurrent tasks.
    b. It tasks are independent, can run in parallel on individual cores
  - **Balance**
    While identifying tasks that can run in parallel, programmers must also ensure that the tasks perform equal work of equal value
  - **Data splitting**
    The data accessed and manipulated by the tasks must be divided to run on separate cores.
  - **Data dependency**
    The data accessed by the tasks must be examined for dependencies between two or more tasks
  - **Testing and debugging**
    When a program is running in parallel on multiple cores, many different execution paths are possible.

Testing and debugging such concurrent programs is inherently more difficult than testing and debugging single-threaded applications.

2.6.2 Types of parallelism

- **Data parallelism**

It distributes subsets of the same data across multiple computing cores and performing the same operation on each core.

Consider, for example, summing the contents of an array of size N.

On a single-core system, one thread would simply sum the elements [0] . . . [N−1].

On a dual-core system, however, thread A, running on core 0, could sum the elements [0] . . . [N/2−1] while thread B, running on core 1, could sum the elements [N/2] ...[N−1]. The two threads would be running in parallel on separate computing cores.

- **Task parallelism**

It distributes tasks (threads) across multiple computing cores. Each thread is performing a unique operation.

Different threads may be operating on the same data, or they may be operating on different data.

An example of task parallelism might involve two threads, each performing a unique statistical operation on the array of elements.

The threads again are operating in parallel on separate computing cores, but each is performing a unique operation.

2.7 MULTITHREADED MODELS

Threads may be used in the user level or kernel level

- User threads - management done by user-level threads library
- Three primary thread libraries:
  a. POSIX P threads
  b. Windows threads
  c. Java threads
- **Kernel threads** - Supported by the Kernel

Examples – virtually all general purpose operating systems, including:

a. Windows

b. Solaris

c. Linux

d. Tru64 UNIX

e. Mac OS X

Multithreading Models

i. Many-to-One

ii. One-to-One

iii. Many-to-Many

i. Many-to-One

- Many user-level threads mapped to single kernel thread
- One thread blocking causes all to block
- Multiple threads may not run in parallel on muticore system because only one may be in kernel at a time
- Few systems currently use this model.

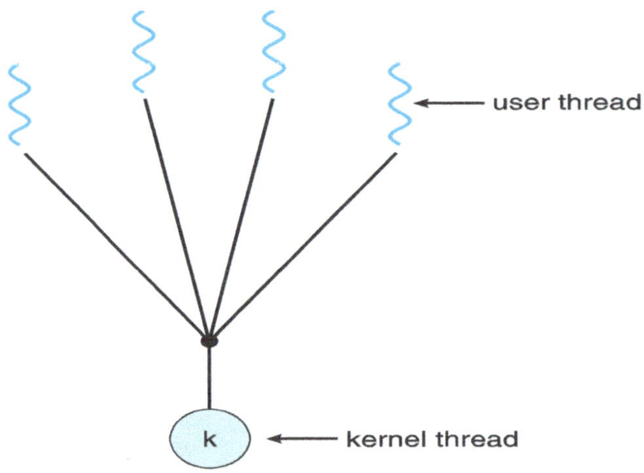

Figure 2.15 Many-to-one model.

- Examples:

  a. Solaris Green Threads

  b. GNU Portable Threads

ii. One-to-One

- Each user-level thread maps to kernel thread
- Creating a user-level thread creates a kernel thread
- More concurrency than many-to-one
- Number of threads per process sometimes restricted due to overhead
- Examples:
    a. Windows
    b. Linux
    c. Solaris 9 and later

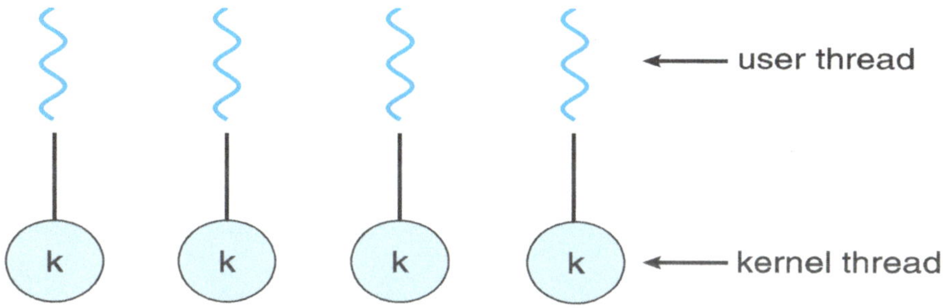

Figure 2.16 One-to-one model.

3. Many-to-Many Model

- Allows many user level threads to be mapped to many kernel threads
- Allows the operating system to create a sufficient number of kernel threads
- Solaris prior to version 9
- Windows with the Thread Fiber package.

Shortcomings

- Developers can create as many user threads as necessary, and the corresponding
  kernel threads can run in parallel on a multiprocessor
- When a thread performs a blocking system call, the kernel can schedule another thread for execution

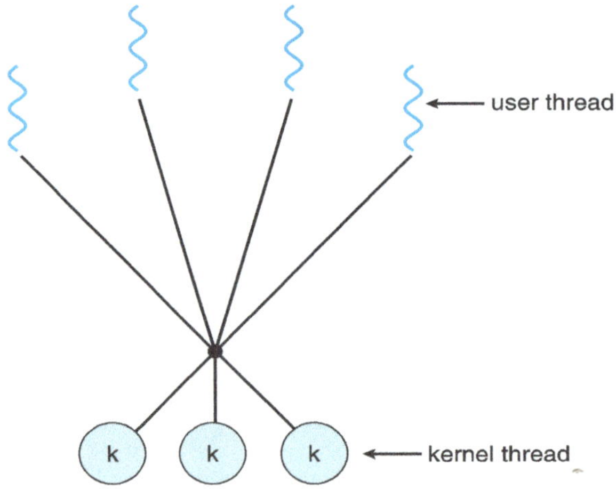

Figure 2.17 Many-to-many model.

Two-level Model

- Similar to Many to many, except that it allows a user thread to be bound to kernel thread
- Examples
  a. IRIX
  b. HP-UX
  c. Tru64 UNIX
  d. Solaris 8 and earlier

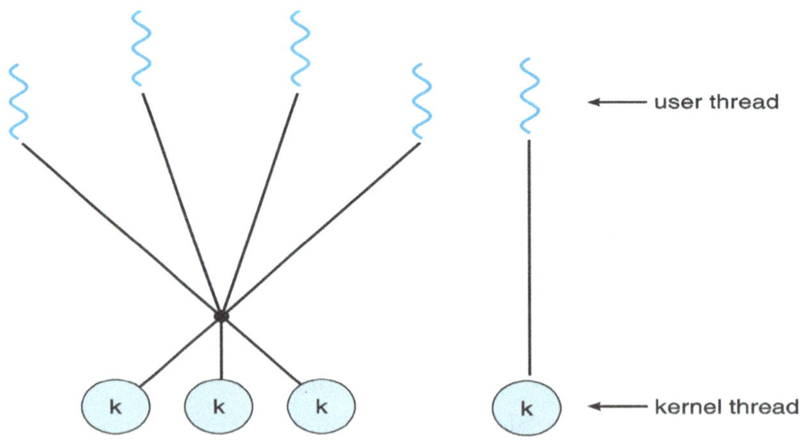

Figure 2.18 Two-level model

2.8 WINDOWS 7 - THREAD AND SMP MANAGEMENT

Windows process design is driven by the need to provide support for a variety of OS environments.

Processes supported by different OS environments

Windows Processes - Important characteristics

Processes and services provided by the Windows Kernel are relatively simple and general purpose.

- Windows processes are implemented as objects.
- A process can be created as new process, or as a copy of an existing process.
- An executable process may contain one or more threads.
- Both process and thread objects have built-in synchronization capabilities.

2.8.1 Relationship between Process and Resources

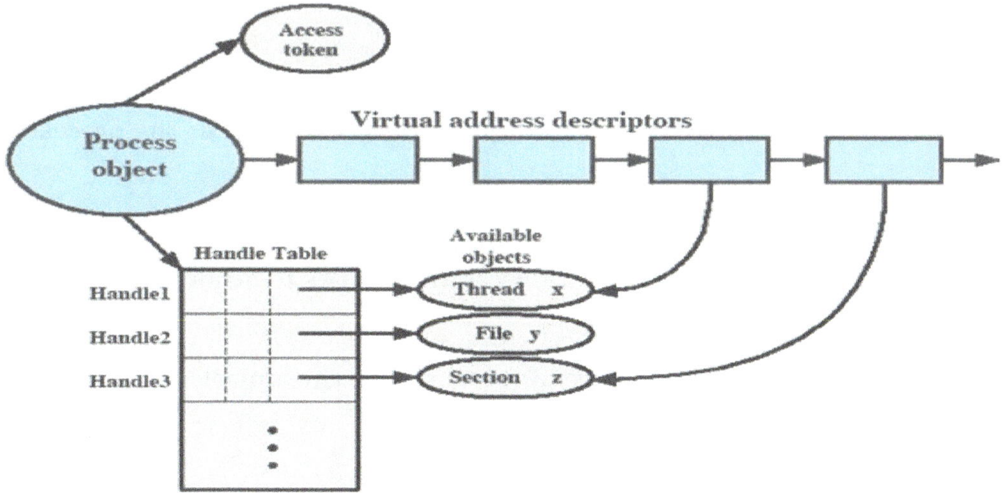

Figure 2.19 A Windows Process and Its Resources

In addition, the process has access to a file object and to a section object that defines a section of shared memory.

Token

- Each process is assigned a security access token, called the primary token of the process.
- When a user first logs on, Windows creates an access token that includes the security ID for the user. Every process that is created by or runs on behalf of this user has a copy of this access token.
- The token is used by windows to validate the user's ability to access secured objects or to perform restricted functions on the system and on secured objects.
- The access token controls whether the process can change its own attributes.
- If the process attempts to open such a handle, the security system determines whether this is permitted and therefore whether the process may change its own attributes.

Address Space

A series of blocks that define the virtual address space currently assigned to this process.

The process cannot directly modify these structures but must rely on the virtual memory manager, which provides a memory allocation service for the process.

Object Table

The process includes an object table, with handles to other objects known to this process.

One handle exists for each thread contained in this object.

2.8.2 Process and Thread Objects

Windows makes use of two types of process-related objects

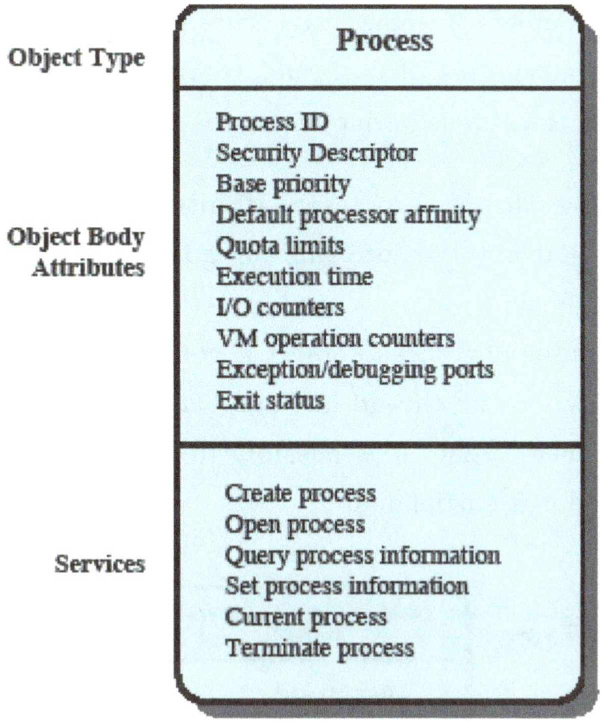

(a) Process object

Figure 2.20a Windows Process object

Processes - an entity corresponding to a user job or application that owns resources

Threads- a dispatchable unit of work that executes sequentially and is interruptible

Windows Process Object:

Each Windows process is represented by an object

Windows makes use of two types of process-related objects: processes and threads.

- A process corresponds to a user job or application that owns resources, such as memory, and opens files.
- A thread is a dispatchable unit of work that executes sequentially and is interruptible, so that the processor can turn to another thread.
  1. A Windows process must contain at least one thread to execute, and the thread may create other threads.
  2. In a multiprocessor system, multiple threads from the same process may execute in parallel.

3. Figure 2.18b depicts the object structure for a thread object
4. Some of the attributes of a thread resemble those of a process. .i.e., the thread attribute value is derived from the process attribute value.

For example, the thread processor affinity is the set of processors in a multiprocessor system that may execute this thread; this set is equal to or a subset of the process processor affinity.

One of the attributes of a thread object is context, which contains the values of the processor registers when the thread last ran. This information enables threads to be suspended and resumed. Also, it is possible to alter the behavior of a thread by altering its context while it is suspended.

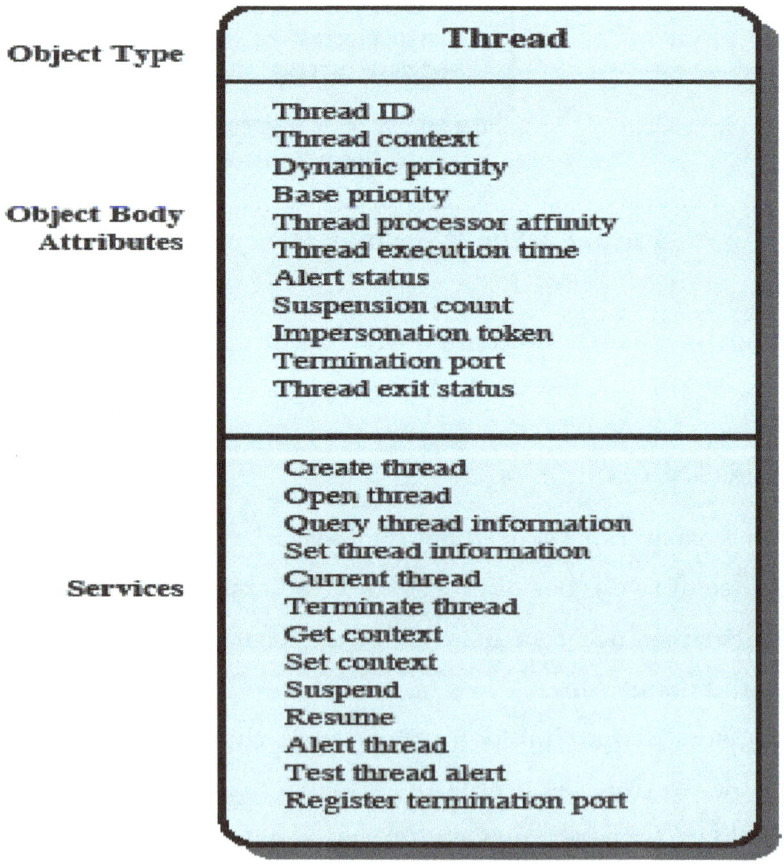

(b) Thread object

Figure 2.20b Windows Thread objects

Multithreaded Process

Achieves concurrency without the overhead of using multiple processes

- Threads within the same process can exchange information through their common address space and have access to the shared resources of the process.
- Threads in different processes can exchange information through shared memory that has been set up between the two processes.

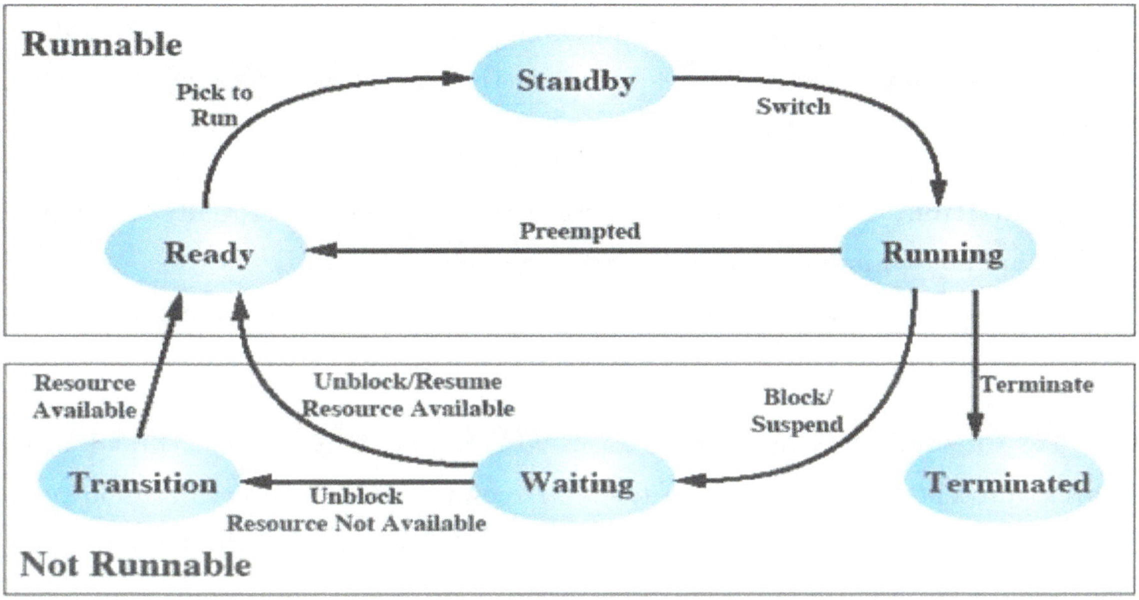

Figure 2.21 Windows Thread States

2.8.3 Thread States

Ready: A ready thread may be scheduled for execution. The Kernel dispatcher keeps track of all ready threads and schedules them in priority order.

Standby: A standby thread has been selected to run next on a particular processor. The thread waits in this state until that processor is made available. If the standby thread's priority is high enough, the running thread on that processor may be preempted in favor of the standby thread. Otherwise, the standby thread waits until the running thread blocks or exhausts its time slice.

Running: Once the Kernel dispatcher performs a thread switch, the standby thread enters the Running state and begins execution and continues execution until it is preempted by a higher-priority thread, exhausts its time slice, blocks, or terminates. In the first two cases, it goes back to the Ready state.

Waiting: A thread enters the Waiting state when

a. It is blocked on an event (e.g., I/O),
b. It voluntarily waits for synchronization purposes, or
c. An environment subsystem directs the thread to suspend itself.

When the waiting condition is satisfied, the thread moves to the Ready state if all of its resources are available.

Transition: A thread enters this state after waiting if it is ready to run but the resources are not available.

For example, the thread's stack may be paged out of memory. When the resources are available, the thread goes to the Ready state.

Terminated: A thread can be terminated by itself, by another thread, or when its parent process terminates. Once tasks are completed, the thread is removed from the system, or it may be retained by the Executive for future re- initialization.

Windows Symmetric Multiprocessing SMP (Support)

- Threads can run on any processor
- But an application can restrict affinity
- In the absence of affinity restrictions, the kernel dispatcher assigns a ready thread to the next available processor. This assures that no processor is idle or is executing a lower-priority thread when a higher-priority thread is ready
- Threads in different processes can exchange information through shared memory that has been set up between the two processes.

Soft Affinity

a. the dispatcher tries to assign a ready thread to the same processor it last ran on

b. helps reuse data still in that processor's memory caches from the previous execution of the thread.

Hard Affinity

a. an application restricts thread execution to certain processors

2.9 PROCESS SYNCHRONIZATION

Background

Concurrent access to shared data may result in data inconsistency.

There are various mechanisms to ensure the orderly execution of cooperating processes that share a logical address space, so that data consistency is maintained.

- Processes can execute concurrently
  - ➢ May be interrupted at any time, partially completing execution
- Concurrent access to shared data may result in data inconsistency
- Maintaining data consistency requires mechanisms to ensure the orderly execution of cooperating processes
- A cooperating process is one that can affect or be affected by other processes executing in the system.
- Ex: Producer Consumer Problem
- Cooperating processes can either directly share a logical address space (that is, both code and data) or be allowed to share data only through files or messages

Race Condition

A situation in which several processes access and manipulate the same data concurrently and the outcome of the execution depends on the particular order in which the access takes place is called race condition.

2.9.1 CRITICAL SECTION PROBLEM

- Consider system of n processes $\{p_0, p_1, \ldots p_{n-1}\}$
- Each process has critical section segment of code
    a. Process may be changing common variables, updating table, writing file, etc
    b. When one process in critical section, no other may be in its critical section
- Critical section problem is to design protocol to solve this
- Each process must ask permission to enter critical section in entry section, may follow critical section with exit section, then remainder section

Critical Section

General structure of process $\mathbf{P_i}$

```
do {
    entry section
        critical section
    exit section
        remainder section
} while (true);
```

Requirements for the solution of Critical-Section Problem

1. **Mutual Exclusion** - If process P_i is executing in its critical section, then no other processes can be executing in their critical sections
2. **Progress** - If no process is executing in its critical section and there exist some processes that wish to enter their critical section, then the selection of the processes that will enter the critical section next cannot be postponed indefinitely

3. **Bounded Waiting** - A bound must exist on the number of times that other processes are allowed to enter their critical sections after a process has made a request to enter its critical section and before that request is granted
    - Assume that each process executes at a nonzero speed
    - No assumption concerning relative speed of the n processes.

Critical-Section Handling in OS

Two approaches depending on if kernel is preemptive or non- preemptive

1. **Preemptive** – allows preemption of process when running in kernel mode
2. **Non-preemptive** – runs until exits kernel mode, blocks, or voluntarily yields CPU
    - Essentially free of race conditions in kernel mode.

Two process Solutions:

Peterson's Solution:

- Good algorithmic description of solving the problem
- Two process solution
    - ➢ Assume that the load and store machine-language instructions are atomic; that is, cannot be interrupted
    - ➢ The two processes share two variables:
        a. int turn;
        b. Boolean flag[2]
        c. The variable turn indicates whose turn it is to enter the critical section
    - ➢ The flag array is used to indicate if a process is ready to enter the critical section. flag[i] = true implies that process P_i is ready!

Algorithm for Process P_i

```
do {
        flag[i] = true;
        turn = j;
        while (flag[j] && turn = = j);
```

```
            critical section
        flag[i] = false;
            remainder section
    } while (true);
```

Provable that the three CS requirement are met:

1. Mutual exclusion is preserved

   P_i enters CS only if:

   either flag[j] = false or turn = i

2. Progress requirement is satisfied
3. Bounded-waiting requirement is met

Synchronization Hardware

- Many systems provide hardware support for implementing the critical section code.
- All solutions below based on idea of locking
  - ➢ Protecting critical regions via locks
- Uniprocessors – could disable interrupts
  - ➢ Currently running code would execute without preemption
  - ➢ Generally too inefficient on multiprocessor systems
    - ❖ Operating systems using this not broadly scalable
- Modern machines provide special atomic hardware instructions
  - ➢ Atomic = non-interruptible
  - ❖ Either test memory word and set value
  - ❖ Or swap contents of two memory words

Solution to Critical-section Problem Using Locks

```
do {
        acquire lock
            critical section
        release lock
            remainder section
```

```
} while (TRUE);
```

Test_and_set Instruction

Definition:

```
boolean test_and_set (boolean *target)
  {
     boolean rv = *target;
     *target = TRUE;
     return rv:
  }
```

1. ecuted atomically
2. Returns the original value of passed parameter
3. Set the new value of passed parameter to "TRUE".

Compare_and_swap Instruction

Definition:

```
int compare _and_swap(int *value, int expected, int new_value) {
   int temp = *value;
   if (*value == expected)
     *value = new_value;
 return temp;
}
```

1. Executed atomically
2. Returns the original value of passed parameter "value"
3. Set the variable "value" the value of the passed parameter "new_value" but only if "value" =="expected". That is, the swap takes place only under this condition.

Using the hardware solutions for the critical section problem the following requirements have met

1. Mutual exclusion is preserved
2. Progress requirement is satisfied
3. Bounded-waiting requirement is met

2.9.2 Mutex

Mutex Locks

The hardware-based solutions to the critical-section problem. The simplest of these tools is the mutex lock. (In fact, the term mutex is short for mutual exclusion).

The mutex locks are used to protect critical regions and thus prevent race conditions.

There are two functions:

- **The acquire ()** function acquires the lock, a process must acquire the lock before entering a critical section and
- **The release ()** function releases the lock, when it exits the critical section.

A mutex lock has a boolean variable available whose value indicates if the lock is available or not.

- If the lock is available, a call to acquire() succeeds, and the lock is then considered unavailable. A process that attempts to acquire an unavailable lock is blocked until the lock is released.
- But this solution requires busy waiting. This lock therefore called a spinlock.

The definition of acquire () is as follows:

```
acquire(){
while (!available); /* busy wait */
available = false;;
}
```

The definition of release () is as follows:

```
release(){
available
```

Solution to the critical-section problem using mutex locks

```
do {
acquire lock
critical section
release lock
```

```
remainder section
}while (true);
= true;
}
```

Calls to either acquire() or release() must be performed atomically. Thus, mutex locks are often implemented using one of the hardware mechanisms.

2.9.3 SEMAPHORE

A semaphore S is an integer variable that, apart from initialization, is accessed only through two standard atomic operations:

- wait () - originally termed P (from the Dutch proberen, "to test");
- signal () - originally called V (from verhogen,"to increment").

The definition of wait () is as follows:

```
wait(S){
while (S<=0)
; // busy wait
S--;
}
```

The definition of signal () is as follows:

```
signal(S){
S++;
}
```

All modifications to the integer value of the semaphore in the wait ()and signal () operations must be executed indivisibly. ie, when one process modifies the semaphore value, no other process can simultaneously modify that same semaphore value.

Semaphore Usage

Operating systems distinguish between

1. Counting semaphore
2. Binary Semaphore

Counting semaphores

1. The value of a counting semaphore can range over an unrestricted domain ie. finite number of instances.
2. The semaphore is initialized to the number of resources available.
3. When a process use a resource, it performs a wait () operation & when a process releases a resource, it performs a signal () operation.

Binary semaphores

1. The value of a binary semaphore can range only between 0 and 1.
2. Thus, binary semaphores behave similarly to mutex locks.

Semaphores can also be used to solve various synchronization problems.

For example, consider two concurrently running processes:

P1 with a statement S1 and P2 with a statement S2. Suppose that S2 is required to execute only after S1 has completed. Then by letting P1 and P2 sharing a common semaphore synch, initialized to 0.

In process P1, the statements are inserted as
S1;
signal(synch);
In processP2, the statements are inserted as
wait(synch);
S2;

Because synch is initialized to 0, P2 will execute S2 only after P1 has invoked signal(synch), which is after statement S1 has been executed.

Semaphore Implementation:

- It must guarantee that no two processes can execute the wait () and signal () on the same semaphore at the same time.
- The implementation becomes the critical section problem where the wait and signal code are placed in the critical section.
- While a process is in its critical section, any other process that tries to enter its critical section must loop continuously in the call to acquire (). ie. It could now have busy waiting in critical section implementation.
- This type of mutex lock is also called a spin lock because the process "spins" while waiting for the lock to become available. But the implementation code is short.
- There will be little busy waiting if critical section rarely occupied.
- This is not a good solution because the applications may spend lots of time in critical sections.

Semaphore Implementation with no busy waiting:

With each semaphore there is an associated waiting queue.

Each entry in a waiting queue has two data items:

- value (of type integer)
- pointer to next record in the list

Two operations:

- **Block** – place the process invoking the operation on the appropriate waiting queue
- **Wakeup** – remove one of processes in the waiting queue and place it in the ready queue

Problems:

Deadlocks

Deadlocks occur where two or more processes are waiting indefinitely for an event that can be caused by one of the waiting processes. When such a state is reached, these processes are said to be deadlocked.

The event is the execution of a signal () operation.

Example

Consider a system consisting of two processes, P0 and P1, each accessing two semaphores, S and Q, set to the value 1:

```
P0             P1
wait(S);       wait(Q);
wait(Q);       wait(S);
  ..             ..
  ..             ..
  ..             ..
signal(S);     signal(Q);
signal(Q);     signal(S);
```

Suppose that P0 executes wait(S) and then P1 executes wait (Q). When P0 executes wait (Q), it must wait until P1 executes signal (Q).
Similarly, when P1 executes wait (S), it must wait until P0 executes signal (S).
Since these signal () operations cannot be executed, P0 and P1are deadlocked.

Indefinite blocking or starvation:

A situation in which processes wait indefinitely within the semaphore and if the processes are removed from the list associated with a semaphore in LIFO (last-in, first-out) order, then starvation occurs.

Classical Problems of Synchronization:

Classical problems used to test newly-proposed synchronization schemes

1. Bounded-Buffer Problem
2. Readers and Writers Problem
3. Dining-Philosophers Problem

2.9.4 Monitors

An abstract data type—or ADT—encapsulates data with a set of functions to operate on that data that are independent of any specific implementation of the ADT.

A monitor type is an ADT that includes a set of programmer defined operations that are provided with mutual exclusion within the monitor.

The monitor type also declares the variables along with the functions.

Thus, a function defined within a monitor can access only those variables declared locally within the monitor and its formal parameters.

Similarly, the local variables of a monitor can be accessed by only the local functions.

The monitor construct ensures that only one process at a time is active within the monitor. But is not sufficiently powerful.

So an additional synchronization mechanisms are provided by the condition construct.

Syntax of a monitor

```
Monitor monitor name
        {
        /* shared variable declarations */
                function P1(...){...}
                function P2(...){...}
                ...
                function Pn(...){...}
                initializationcode(...){...}
        }
```

Condition Variable:

The only operations that can be invoked on a condition variable are

1. wait()
2. signal().

- The operation x.wait ()- the process is suspended until another process invokes.
- The operation x.signal ()- resumes exactly one suspended process.

If no process is suspended, then the signal () operation has no effect;

When the x.signal () operation is invoked by a process P, there exists a suspended process Q associated with condition x.

If the suspended process Q is allowed to resume its execution, then the signaling process P must wait.

Otherwise, both P and Q would be active simultaneously within the monitor.

But both processes can continue with their execution.

Two possibilities exist:

1. **Signal and wait:** P either waits until Q leaves the monitor or waits for another condition.
2. **Signal and continue:** Q either waits until P leaves the monitor or waits for another condition.

Operations queues associated with x, y conditions.

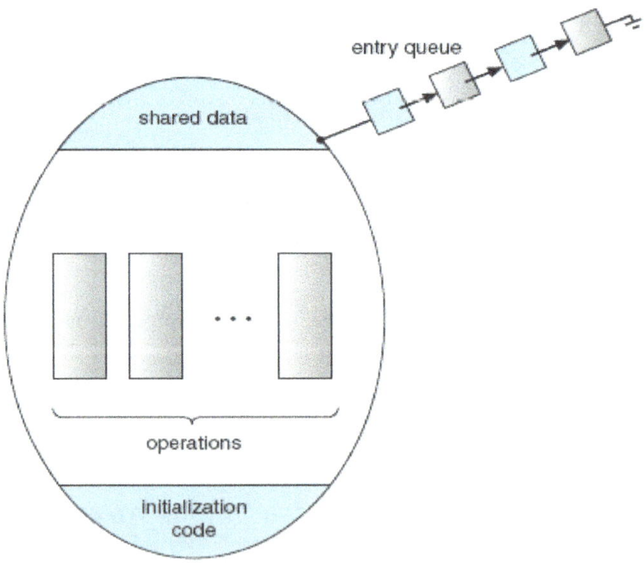

Figure 2.22 Schematic view of a monitor

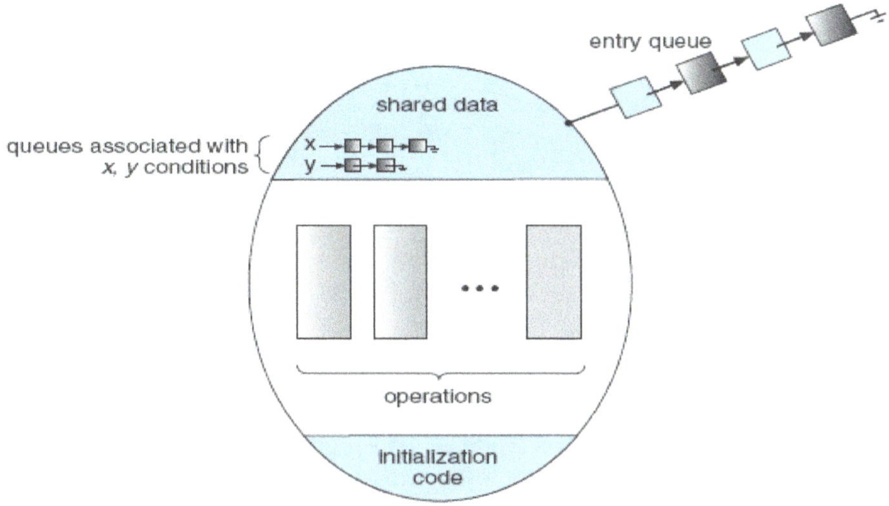

Figure 2.23 Monitor with condition variables

There are reasons in favor of adopting either option.

- Since P was already executing in the monitor, the signal-and continue method seems more reasonable.
- If thread P is allowed to continue, then by the time Q is resumed, the logical condition for which Q was waiting may no longer hold.

Monitors implemented in the language Concurrent Pascal.

- When thread P executes the signal operation, it immediately leaves the monitor. Hence, Q is immediately resumed.

Implemented in other languages including Mesa, C#, Java.

Dining-Philosophers Problem:

- Philosophers spend their lives alternating thinking and eating
- Don't interact with their neighbors, occasionally try to pick up 2 chopsticks (one at a time) to eat from bowl
  - ➢ Need both to eat, then release both when done
- In the case of 5 philosophers
  - ➢ Shared data
    - ❖ Bowl of rice (data set)
    - ❖ Semaphore chopstick [5] initialized to 1

Figure 2.24 Dining-Philosophers

The structure of Philosopher i:

```
do {
      wait (chopstick[i] );
            wait (chopStick[ (i + 1) % 5] );
                        // eat
            signal (chopstick[i] );
            signal (chopstick[ (i + 1) % 5] );

            // think
} while (TRUE);
```

Problem with algorithm:

Incorrect use of semaphore operations:

- signal (mutex) .... wait (mutex)
- wait (mutex) ... wait (mutex)
- Omitting of wait (mutex) or signal (mutex) (or both)
- Deadlock and starvation.

Dining-Philosophers Solution Using Monitors:

```
monitor DiningPhilosophers
{
enum{THINKING, HUNGRY, EATING} state[5];
condition self[5];
void pickup(int i) {
state[i] = HUNGRY;
test(i);
if (state[i] != EATING)
self[i].wait();
}
void putdown(int i) {
state[i] = THINKING;
test((i + 4) % 5);
test((i + 1) % 5);
}
void test(int i) {
if ((state[(i + 4) % 5] != EATING) &&
(state[i] == HUNGRY) &&
(state[(i + 1) % 5] != EATING)){
state[i] = EATING;
self[i].signal();
}
}
initializationcode() {
for(inti=0;i<5;i++)
state[i] = THINKING;
}
}
```

Each philosopher i must invoke the operations pickup() and putdown() in the following sequence:

```
DiningPhilosophers.pickup(i);
...
eat
...
DiningPhilosophers.putdown(i);
```

No deadlock, but starvation is possible.

Monitor Implementation Using Semaphores

Variables

```
semaphore mutex; // (initially = 1)
semaphore next;  // (initially = 0)
int next_count = 0;
```

Each procedure F will be replaced by

```
wait(mutex);
        ...
 body of F;
        ...
if (next_count > 0)
        signal(next)
else
        signal(mutex);
```

Mutual exclusion within a monitor is ensured

Monitor Implementation – Condition Variables

For each condition variable x, we have:

```
semaphore x_sem; // (initially = 0)
int x_count = 0;
```

The operation x.wait can be implemented as:

```
x_count++;
if (next_count > 0)
        signal(next);
else
        signal(mutex);
wait(x_sem);
x_count--;
```

The operation x.signal can be implemented as:

```
        if (x_count > 0) {
        next_count++;
        signal(x_sem);
        wait(next);
        next_count--;
}
```

Resuming Processes within a Monitor

- If several processes queued on condition x, and x.signal () executed, which should be resumed?
- FCFS frequently not adequate
- conditional-wait construct of the form x.wait(c)
  - ❖ Where c is priority number
  - ❖ Process with lowest number (highest priority) is scheduled next

Single Resource allocation

Allocate a single resource among competing processes using priority numbers that specify the maximum time a process plans to use the resource.

```
R.acquire(t);
  ...
 access the resource;
```

```
    ...
  R.release;
```

Where R is an instance of type ResourceAllocator

A Monitor to Allocate Single Resource

```
monitor ResourceAllocator
{
        boolean busy;
        condition x;
        void acquire(int time) {
                if (busy)
                        x.wait(time);
                busy = TRUE;
        }
        void release() {
                busy = FALSE;
                x.signal();
        }
initialization code() {
         busy = FALSE;
        }
}
```

2.10 CPU Scheduling

1. Basic Concepts

- Maximum CPU utilization obtained with multiprogramming
- CPU–I/O Burst Cycle – Process execution consists of a cycle of CPU execution and I/O wait
- CPU burst followed by I/O burst
- CPU burst distribution is of main concern

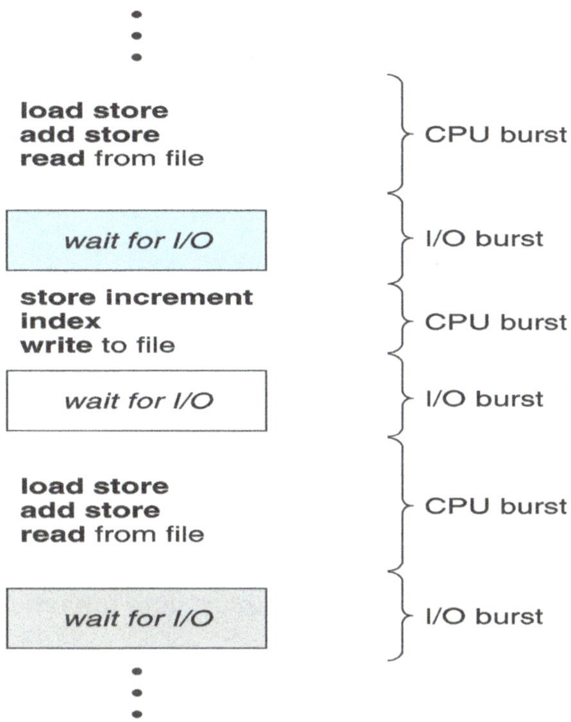

Figure 2.25 Alternating sequence of CPU and I/O bursts.

CPU Scheduler

- **Short-term scheduler** selects from among the processes in ready queue, and allocates the CPU to one of them
    - ➢ Queue may be ordered in various ways
- CPU scheduling decisions may take place when a process:
    - ➢ Switches from running to waiting state
    - ➢ Switches from running to ready state
- Switches from waiting to ready
- Terminates
- Scheduling under 1 and 4 is non preemptive
- All other scheduling is preemptive
    - ➢ Consider access to shared data
    - ➢ Consider preemption while in kernel mode
    - ➢ Consider interrupts occurring during crucial OS activities

Dispatcher

- Dispatcher module gives control of the CPU to the process selected by the short-term scheduler; this involves:
  - switching context
  - switching to user mode
  - jumping to the proper location in the user program to restart that program
- **Dispatch latency** – time it takes for the dispatcher to stop one process and start another running

Scheduling Criteria

- **CPU utilization** – keep the CPU as busy as possible
- **Throughput** – # of processes that complete their execution per time unit
- **Turnaround time** – amount of time to execute a particular process
- **Waiting time** – amount of time a process has been waiting in the ready queue
- **Response time** – amount of time it takes from when a request was submitted until the first response is produced, not output (for time-sharing environment)

Scheduling Algorithm Optimization Criteria

1. Max CPU utilization
2. Max throughput
3. Min turnaround time
4. Min waiting time
5. Min response time

2.10.1 Scheduling Algorithms

1. First Come, First-Served Scheduling (FCFS)
2. Shortest-Job-First (SJF) Scheduling
3. Priority Scheduling
4. Round Robin (RR)
5. Multilevel Queue
6. Multilevel Feedback Queue

1. First- Come, First-Served (FCFS) Scheduling

In this algorithm, the process that requests the CPU first is allocated the CPU first. This implementation of the FCFS policy is easily managed with a FIFO queue.

Process	Burst Time
P_1	24
P_2	3
P_3	3

Suppose that the processes arrive in the order: P_1 , P_2 , P_3 The Gantt Chart for the schedule is:

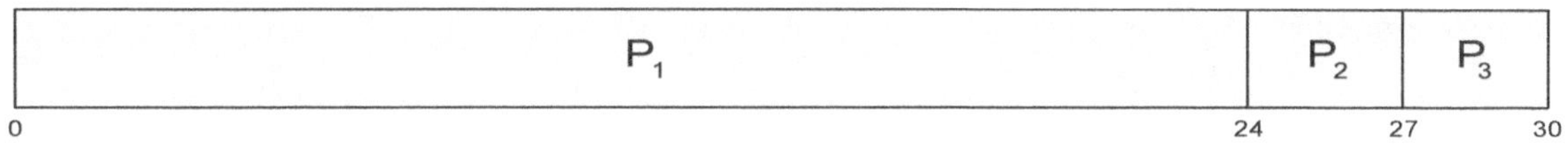

Waiting time for $P_1 = 0$; $P_2 = 24$; $P_3 = 27$

Average waiting time: $(0 + 24 + 27)/3 = 17$

- **Suppose that the processes arrive in the order**:

  P_2 , P_3 , P_1

  The Gantt chart for the schedule is:

- Waiting time for $P_1 = 6$; $P_2 = 0$; $P_3 = 3$
- Average waiting time: $(6 + 0 + 3)/3 = 3$
- Much better than previous case
- **Convoy effect** - short process behind long process

  Consider one CPU-bound and many I/O-bound processes

2. Shortest-Job-First (SJF) Scheduling

This algorithm associates with each process the length of the process's next CPU burst. When the CPU is available, it is assigned to the process that has the smallest next CPU burst. If the next CPU bursts of two processes are the same, FCFS scheduling is used to break the tie.

Associate with each process the length of its next CPU burst

- Use these lengths to schedule the process with the shortest time

SJF is optimal – gives minimum average waiting time for a given set of processes

- The difficulty is knowing the length of the next CPU request
- Could ask the user

Example of SJF

Process	Arrival Time	Burst Time
P_1	0.0	6
P_2	2.0	8
P_3	4.0	7
P_4	5.0	3

SJF scheduling chart

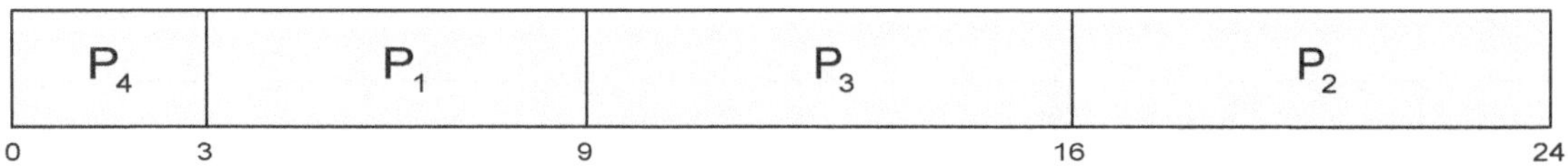

Average waiting time = (3 + 16 + 9 + 0) / 4 = 7

Determining Length of Next CPU Burst

- Can only estimate the length – should be similar to the previous one
  - ➢ Then pick process with shortest predicted next CPU burst
- Can be done by using the length of previous CPU bursts, using exponential averaging
- Preemptive version called shortest-remaining-time-first (SRF)

Example of Shortest-remaining-time-first (SRF)

Now we add the concepts of varying arrival times and preemption to the analysis

Process	Arrival Time	Burst Time
P_1	0	8
P_2	1	4
P_3	2	9
P_4	3	5

Preemptive SJF Gantt Chart

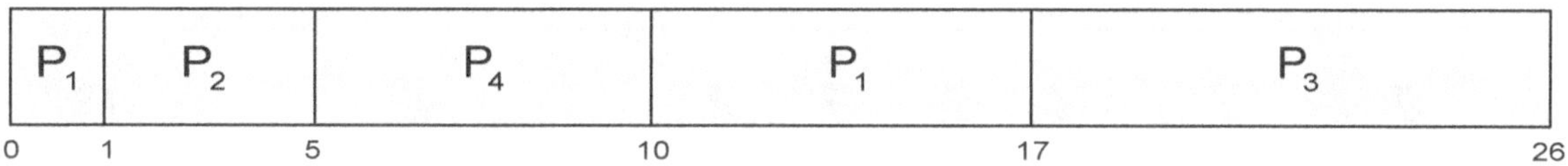

Average waiting time = [(10-1)+(1-1)+(17-2)+5-3)]/4 = 26/4 = 6.5 msec

EXAMPLES

i. Normal SJF

Process	Burst Time
P_1	7
P_2	3
P_3	4

The Gantt Chart for SJF (Normal) is:

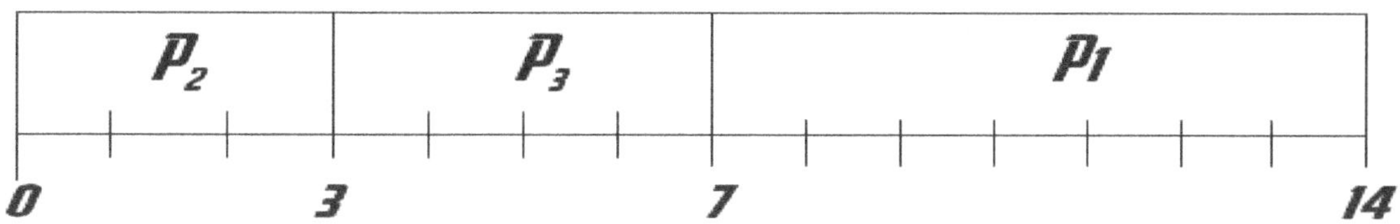

Average waiting time = (0 + 3 + 7)/3 = 3.33

ii. Non-Preemptive SJF

Process	Arrival Time	Burst Time
P_1	0.0	7
P_2	2.0	4
P_3	4.0	1
P_4	5.0	4

The Gantt Chart for SJF (non-preemptive) is:

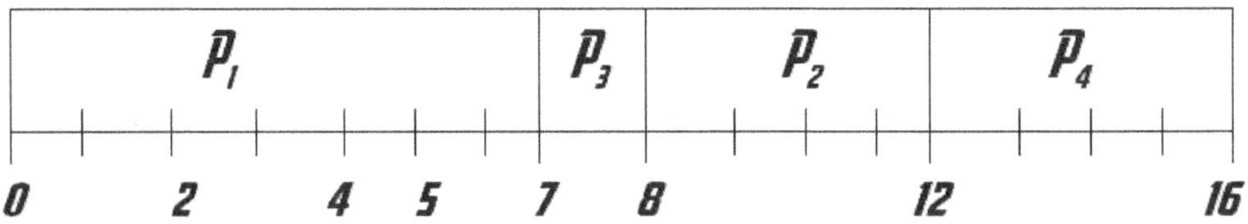

Average waiting time = (0 + 6 + 3 + 7)/4 = 4

iii. Preemptive SJF

Process	Arrival Time	Burst Time
P1	0.0	7
P2	2.0	4
P3	4.0	1
P4	5.0	4

The Gantt Chart for SJF (preemptive) is:

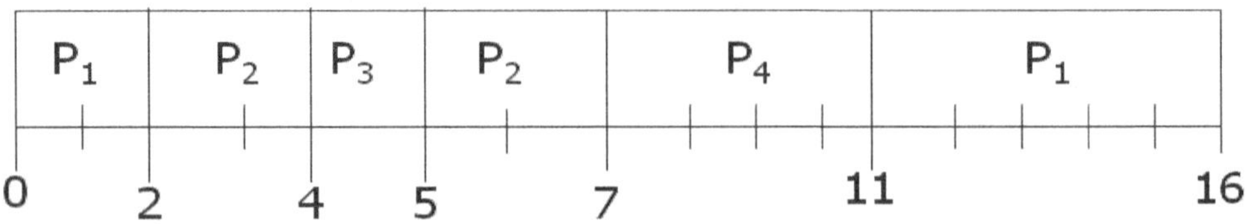

Average waiting time = (9 + 1 + 0 +2)/4 = 3

Example 3:

Process	Arrival Time	Burst Time
P_1	0.0	6
P_2	0.0	4
P_3	0.0	1
P_4	0.0	5

i) SJF (non-preemptive, simultaneous arrival)

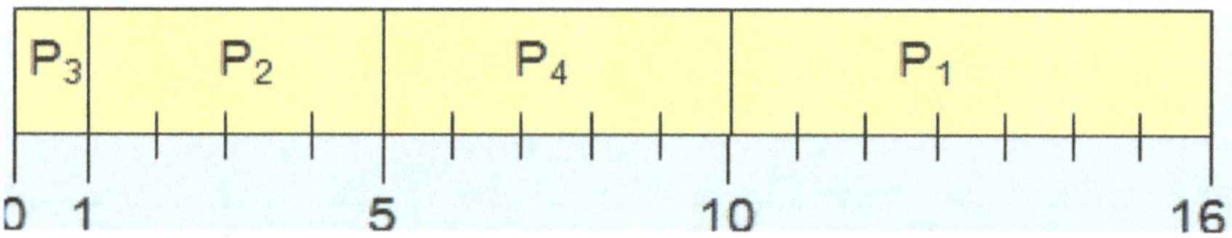

Average waiting time = (0 + 1 + 5 + 10)/4 = 4

Average turn-around time = (1 + 5 + 10 + 16)/4 = 8

ii) Non-Preemptive SJF (varied arrival times)

Process	Arrival Time	Burst Time
P_1	0.0	7
P_2	2.0	4
P_3	4.0	1
P_4	5.0	4

iii) SJF (non-preemptive, varied arrival times)

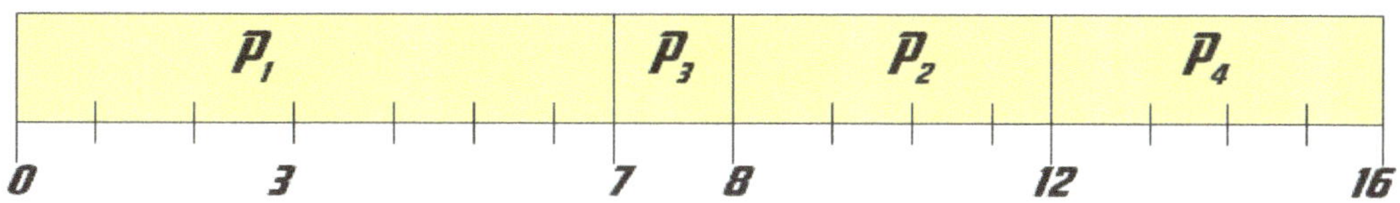

Average waiting time = ((0 – 0) + (8 – 2) + (7 – 4) + (12 – 5))/4
= (0 + 6 + 3 + 7)/4 = 4

Average turn-around time = ((7 – 0) + (12 – 2) + (8 - 4) + (16 – 5))/4
= (7 + 10 + 4 + 11) / 4 = 8

iv) Pre-emptive SJF (Shortest-remaining-time-first)

Process	Arrival Time	Burst Time
P_1	0.0	7
P_2	2.0	4
P_3	4.0	1
P_4	5.0	4

SJF (pre-emptive, varied arrival times)

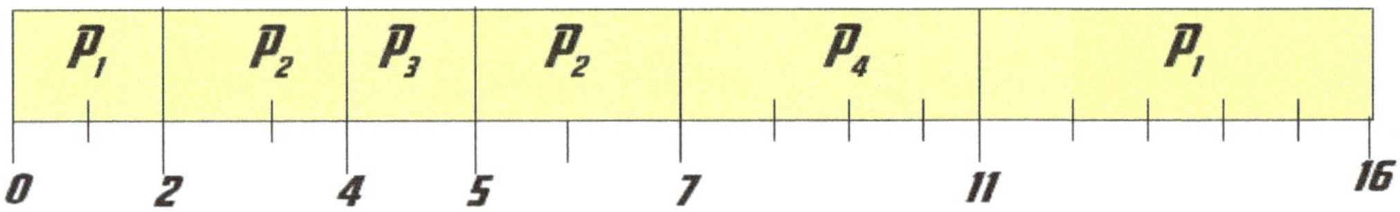

Average waiting time = ([(0 – 0) + (11 - 2)] + [(2 – 2) + (5 – 4)] + (4 - 4) + (7 – 5))/4
= 9 + 1 + 0 + 2)/4
= 3

Average turn-around time = (16 + 7 + 5 + 11)/4 = 9.75

3. Priority Scheduling

- A priority number (integer) is associated with each process
- The CPU is allocated to the process with the highest priority (smallest integer ≡ highest priority)
  - Preemptive.
  - Non preemptive.
  - SJF is priority scheduling where priority is the inverse of predicted next CPU burst time.
- Problem ≡ **Starvation** – low priority processes may never execute
- Solution ≡ **Aging** – as time progresses increase the priority of the process

Example 1

Process	Burst Time	Priority
P_1	10	3
P_2	1	1
P_3	2	4
P_4	1	5
P_5	5	2

Priority scheduling Gantt Chart

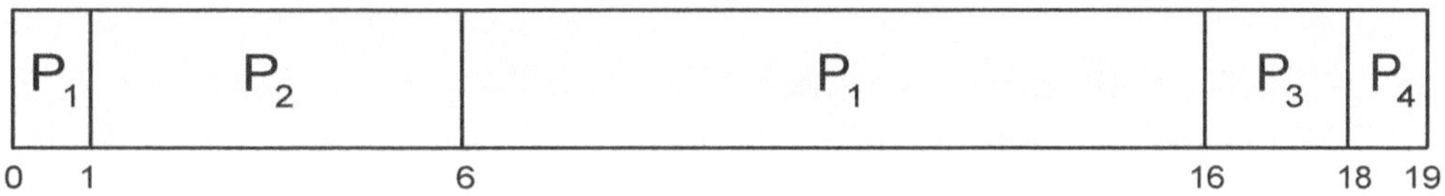

Average waiting time = 8.2 msec

Example 2

Process	Burst Time	Priority
P1	10	3
P2	1	1
P3	2	4
P4	1	5
P5	5	2

Gantt Chart

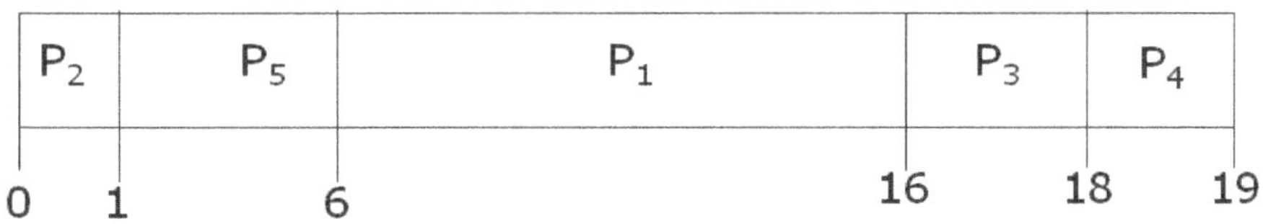

Average waiting time = (6 + 0 + 16 + 18 + 1)/5 = 8.2

4. Round Robin (RR)

- Each process gets a small unit of CPU time (time quantum q), usually 10-100 milliseconds. After this time has elapsed, the process is preempted and added to the end of the ready queue.
- If there are n processes in the ready queue and the time quantum is q, then each process gets 1/n of the CPU time in chunks of at most q time units at once. No process waits more than (n-1)q time units.
- Timer interrupts every quantum to schedule next process
- Performance
  - q large ⇒ FIFO
  - q small ⇒ q must be large with respect to context switch, otherwise overhead is too high

Example 1: RR with Time Quantum = 4

Process	Burst Time
P_1	24
P_2	3
P_3	3

The Gantt chart is:

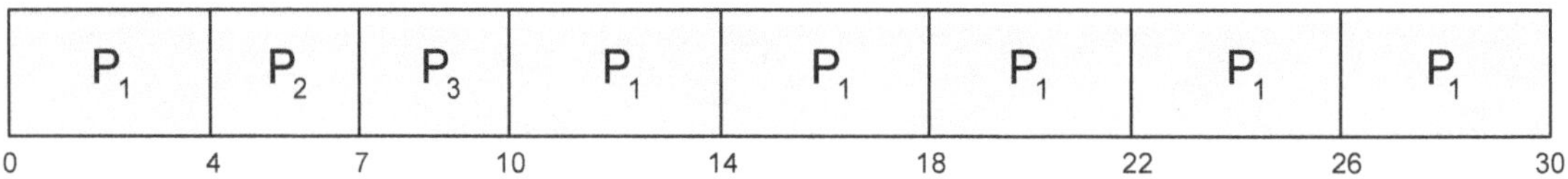

- Typically, higher average turnaround than SJF, but better response
- q should be large compared to context switch time
- q usually 10ms to 100ms, context switch < 10 usec

Example 2: RR with Time Quantum = 20

Process	Burst Time
P_1	53
P_2	17
P_3	68
P_4	24

The Gantt chart is:

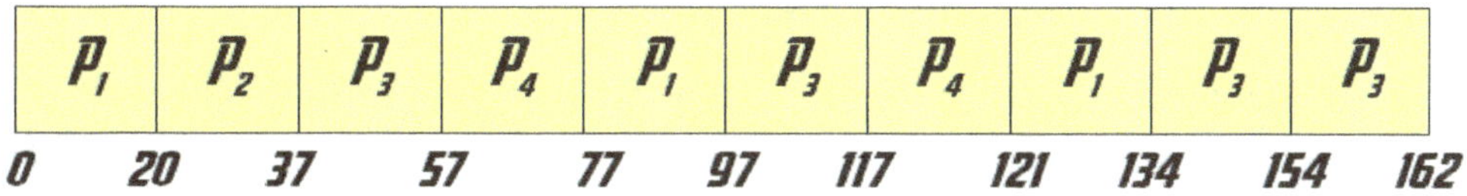

Typically, higher average turnaround than SJF, but better response time

Average waiting time = ([(0 – 0) + (77 - 20) + (121 – 97)] + (20 – 0) + [(37 – 0) + (97 - 57) + (134 – 117)] + [(57 – 0) + (117 – 77)]) / 4

= (0 + 57 + 24) + 20 + (37 + 40 + 17) + (57 + 40)) / 4

= (81 + 20 + 94 + 97)/4

= 292 / 4 = 73

Average turn-around time = 134 + 37 + 162 + 121) / 4 = 113.5

Time Quantum and Context Switch Time

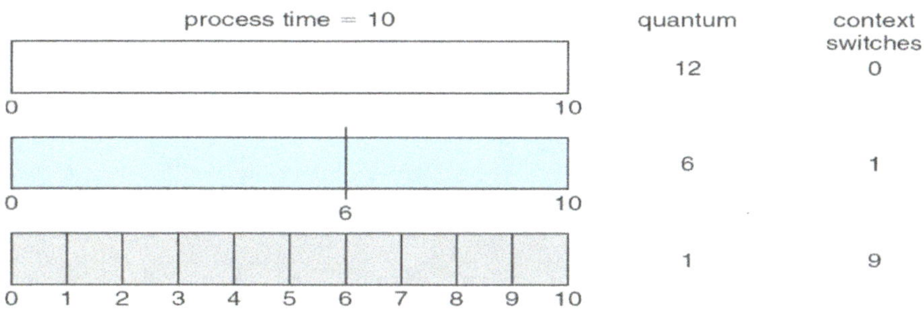

Figure 2.26 How a smaller time quantum increases context switches

2.10.1.5.1 Multilevel Queue Scheduling

- Ready queue is partitioned into separate queues, eg:
  - foreground (interactive)
  - background (batch)
- Process permanently in a given queue
- Each queue has its own scheduling algorithm:
  - foreground – RR
  - background – FCFS
- Scheduling must be done between the queues:
  - Fixed priority scheduling; (i.e., serve all from foreground then from background). Possibility of starvation.
  - Time slice – each queue gets a certain amount of CPU time which it can schedule amongst its processes; i.e., 80% to foreground in RR , 20% to background in FCFS.

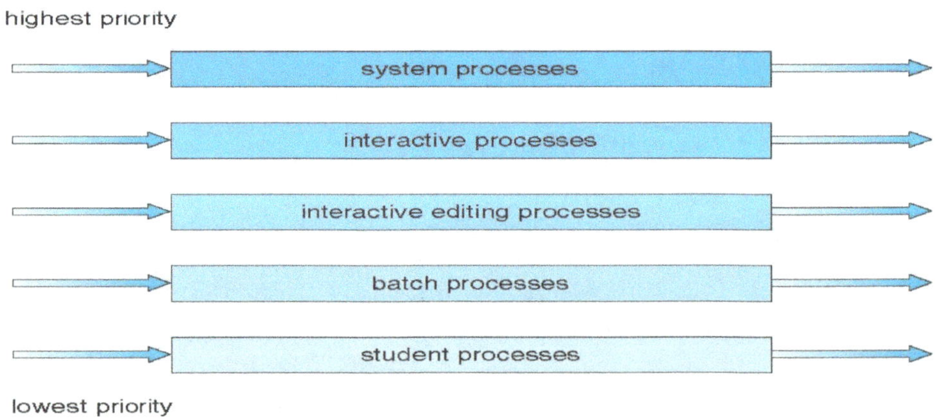

Figure 2.27 Multilevel queue scheduling.

5. Multilevel Feedback Queue

- A process can move between the various queues; aging can be implemented this way
- Multilevel-feedback-queue scheduler defined by the following parameters:
  - number of queues
  - scheduling algorithms for each queue

- method used to determine when to upgrade a process
- method used to determine when to demote a process
- method used to determine which queue a process will enter when that process needs service

- Three queues:
  - Q_0 – RR with time quantum 8 milliseconds
  - Q_1 – RR time quantum 16 milliseconds
  - Q_2 – FCFS
  - Scheduling
  - A new job enters queue Q_0 which is served FCFS
    - When it gains CPU, job receives 8 milliseconds
    - If it does not finish in 8 milliseconds, job is moved to queue Q_1
  - At Q_1 job is again served FCFS and receives 16 additional milliseconds
    - If it still does not complete, it is preempted and moved to queue Q_2

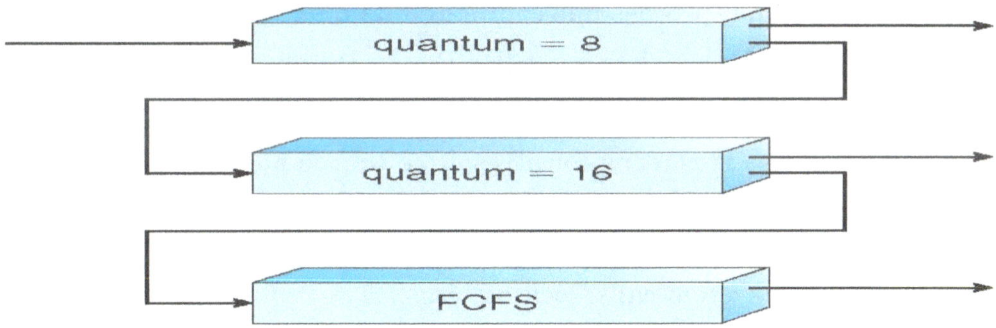

Figure 2.28 Multilevel feedback queues

2.11 DEADLOCKS

In a multiprogramming environment, several processes may compete for a finite number of resources. A process requests resources; if the resources are not available at that time, the process enters a waiting state. Sometimes, a waiting process is never again able to change state, because the resources it has requested are held by other waiting processes. This situation is called a deadlock.

System Model

- System consists of resources
- Resource types $R_1, R_2, \ldots, R_m$
  - CPU cycles, memory space, I/O devices
- Each resource type R_i has W_i instances.
- Each process utilizes a resource as follows:
  - request
  - use
  - release

Deadlock Characterization:

Deadlock can arise if four conditions hold simultaneously.

1. **Mutual exclusion:** only one process at a time can use a resource
2. **Hold and wait:** a process holding at least one resource is waiting to acquire additional resources held by other processes
3. **No preemption:** a resource can be released only voluntarily by the process holding it, after that process has completed its task
4. **Circular wait:** there exists a set $\{P_0, P_1, \ldots, P_n\}$ of waiting processes such that P_0 is waiting for a resource that is held by P_1, P_1 is waiting for a resource that is held by P_2, …, P_{n-1} is waiting for a resource that is held by P_n, and P_n is waiting for a resource that is held by P_0.

Deadlocks can occur via system calls, locking, etc.

Resource-Allocation Graph

A set of vertices V and a set of edges E.

- V is partitioned into two types:
  - $P = \{P_1, P_2, \ldots, P_n\}$, the set consisting of all the processes in the system
  - $R = \{R_1, R_2, \ldots, R_m\}$, the set consisting of all resource types in the system
  - request edge – directed edge $P_i \rightarrow R_j$
- assignment edge – directed edge $R_j \rightarrow P_i$

- Process

- Resource Type with 4 instances

- P_i requests instance of R_j

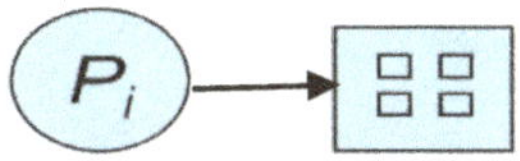

- P_i is holding an instance of R_j

Example of a Resource Allocation Graph

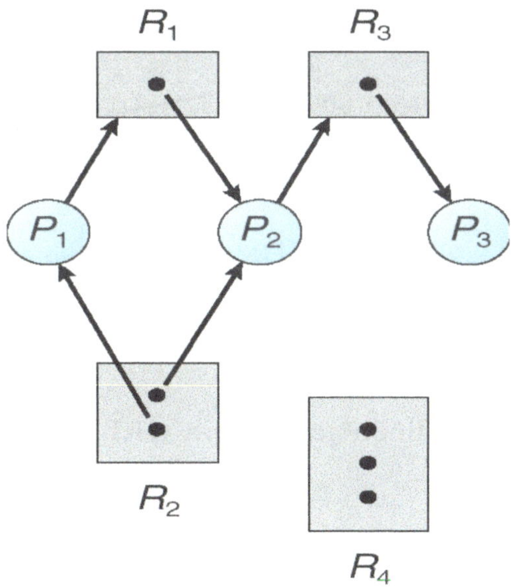

Figure 2.29 Resource-allocation graphs.

Resource Allocation Graph With A Deadlock

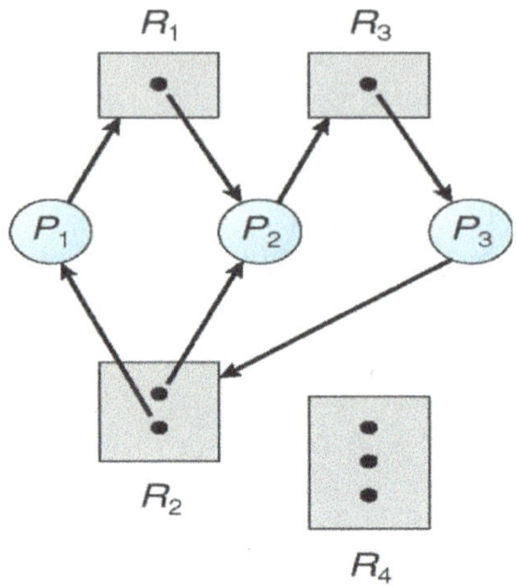

Figure 2.30 Resource-allocation graph with a deadlock.

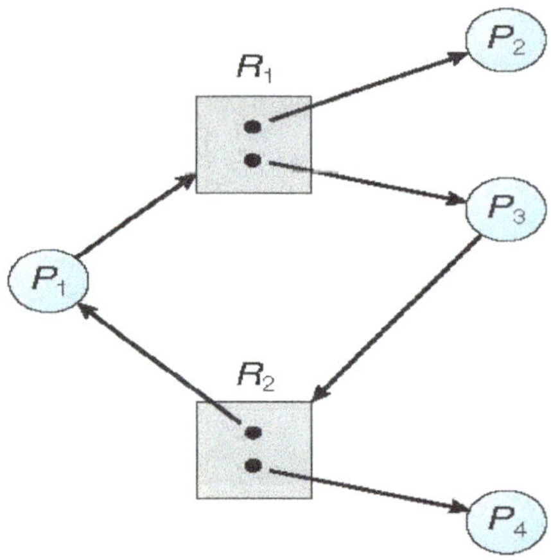

Figure 2.31 Resource-allocation graph with a cycle but no deadlock

Basic Facts

- If graph contains no cycles ⇒ no deadlock
- If graph contains a cycle ⇒
  - if only one instance per resource type, then deadlock
  - if several instances per resource type, possibility of deadlock

2.11.1 Methods for handling Deadlocks

- Ensure that the system will never enter a deadlock state:
    1. Deadlock prevention
    2. Deadlock avoidance
    3. Deadlock detection
    4. Deadlock recovery
- Allow the system to enter a deadlock state and then recover.
- Ignore the problem and pretend that deadlocks never occur in the system; used by most operating systems, including UNIX.

2.11.1.1 Deadlock Prevention

Restrain the ways request can be made

- **Mutual Exclusion** – not required for sharable resources (e.g., read-only files); must hold for non-sharable resources
- **Hold and Wait** – must guarantee that whenever a process requests a resource, it does not hold any other resources
    - Require process to request and be allocated all its resources
      Before it begins execution, or allows process to request resources only when the process has none allocated to it.
    - Low resource utilization; starvation possible
- **No Preemption**
    - If a process that is holding some resources requests another resource that cannot be immediately allocated to it, then all resources currently being held are released
    - Preempted resources are added to the list of resources for which the process is waiting
    - Process will be restarted only when it can regain its old
      resources, as well as the new ones that it is requesting
- **Circular Wait** - impose a total ordering of all resource types, and require that each process requests resources in an increasing order of enumeration

2.11.1.2 Deadlock Avoidance

Requires that the system has some additional a priori information available

- Simplest and most useful model requires that each process declare the maximum number of resources of each type that it may need
- The deadlock-avoidance algorithm dynamically examines the resource-allocation state to ensure that there can never be a circular-wait condition
- Resource-allocation state is defined by the number of available and allocated resources, and the maximum demands of the processes

2.11.1.2.1 Safe State

- When a process requests an available resource, system must decide if immediate allocation leaves the system in a safe state
- System is in safe state if there exists a sequence $<P_1, P_2, \ldots, P_n>$ of ALL the processes in the systems such that for each P_i, the resources that P_i can still request can be satisfied by currently available resources + resources held by all the P_j, with $j < I$
- That is:

➢ If P_i resource needs are not immediately available, then P_i can wait until all P_j have finished

➢ When P_j is finished, P_i can obtain needed resources, execute, return allocated resources, and terminate

➢ When P_i terminates, P_{i+1} can obtain its needed resources, and so on

Basic Facts

- If a system is in safe state ⇒ no deadlocks
- If a system is in unsafe state ⇒ possibility of deadlock
- Avoidance ⇒ ensure that a system will never enter an unsafe state.

  Safe, Unsafe, Deadlock State

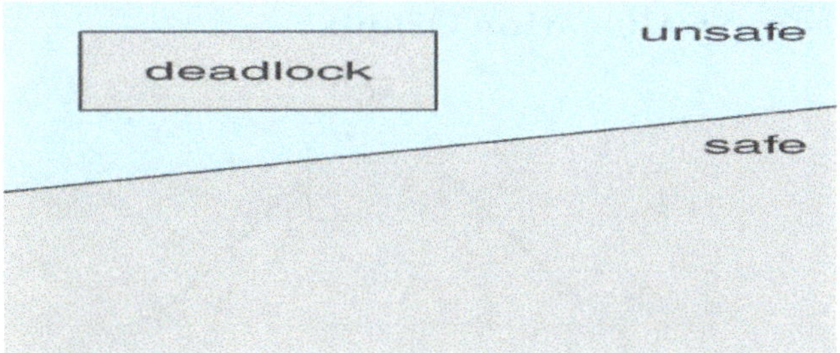

Figure 2.32 Safe, unsafe, and deadlocked state spaces.

2.11.1.2.2 Avoidance Algorithms

- Single instance of a resource type
  - Use a resource-allocation graph
- Multiple instances of a resource type
  - Use the banker's algorithm
    - Safety Algorithm
    - Resource-Request Algorithm

2.11.1.2.2.1 Resource-Allocation Graph Scheme

- Claim edge $P_i \rightarrow R_j$ indicated that process P_j may request resource R_j; represented by a dashed line
- Claim edge converts to request edge when a process requests a resource
- Request edge converted to an assignment edge when the resource is allocated to the process
- When a resource is released by a process, assignment edge reconverts to a claim edge
- Resources must be claimed a priori in the system

Unsafe State in Resource-Allocation Graph

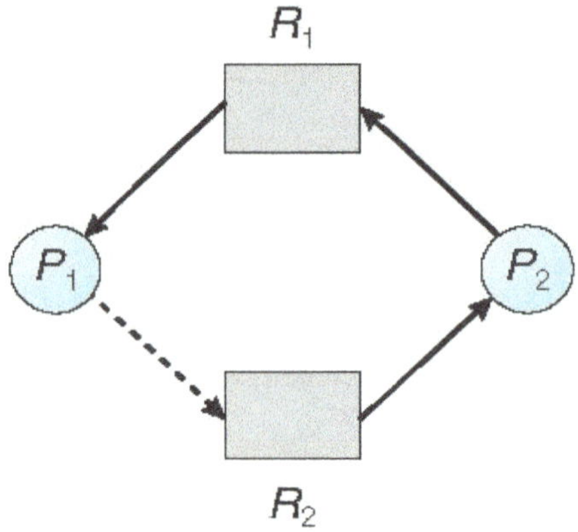

Figure 2.33 An unsafe state in a resource-allocation graph

Resource-Allocation Graph Algorithm

- Suppose that process P_i requests a resource R_j
- The request can be granted only if converting the request edge to an assignment edge does not result in the formation of a cycle in the resource allocation graph

2.11.1.2.2.2. Banker's Algorithm

- Multiple instances
- Each process must a priori claim maximum use
- When a process requests a resource it may have to wait
- When a process gets all its resources it must return them in a finite amount of time

Data Structures for the Banker's Algorithm

Let n = number of processes, and m = number of resources types

1. **Available**: Vector of length m. If available [j] = k, there are k instances of resource type R_j available
2. **Max**: n x m matrix. If Max [i,j] = k, then process P_i may request at most k instances of resource type R_j
3. **Allocation**: n x m matrix. If Allocation[i,j] = k then P_i is currently allocated k instances of R_j

4. **Need**: n x m matrix. If Need[i,j] = k, then P_i may need k more instances of R_j to complete its task

Need [i,j] = Max[i,j] – Allocation [i,j]

2.11.1.2.2.2.1 Safety Algorithm

1. Let Work and Finish be vectors of length m and n, respectively.

   Initialize:

   Work = Available

   Finish [i] = false for i = 0, 1, …, n- 1
2. Find an i such that both:

   (a) Finish [i] = false

   (b) $Need_i \leq Work$

   If no such i exists, go to step 4
3. Work = Work + $Allocation_i$ Finish[i] = true go to step 2
4. If Finish [i] == true for all i, then the system is in a safe state

2.11.1.2.2.2.2 Resource-Request Algorithm for Process P_i

$Request_i$ = request vector for process P_i. If $Request_i$ [j] = k then process P_i wants k instances of resource type R_j

1. If $Request_i \leq Need_i$ go to step 2. Otherwise, raise error condition, since process has exceeded its maximum claim
2. If $Request_i \leq Available$, go to step 3. Otherwise P_i must wait, since resources are not available
3. Pretend to allocate requested resources to P_i by modifying the state as follows:

   Available = Available – $Request_i$;

   $Allocation_i$ = $Allocation_i$ + $Request_i$;

   $Need_i$ = $Need_i$ – $Request_i$;

   - If safe ⇒ the resources are allocated to P_i
   - If unsafe ⇒ P_i must wait, and the old resource-allocation state is restored

Example of Banker's Algorithm

- 5 processes P_0 through P_4;
- 3 resource types:

A (10 instances), B (5instances), and C (7 instances)

- Snapshot at time T_0:

	Allocation	Max	Available
	A B C	A B C	A B C
P_0	0 1 0	7 5 3	3 3 2
P_1	2 0 0	3 2 2	
P_2	3 0 2	9 0 2	
P_3	2 1 1	2 2 2	
P_4	0 0 2	4 3 3	

- The content of the matrix Need is defined to be Max – Allocation

	Need
	A B C
P_0	7 4 3
P_1	1 2 2
P_2	6 0 0
P_3	0 1 1
P_4	4 3 1

- The system is in a safe state since the sequence $< P_1, P_3, P_4, P_2, P_0>$ satisfies safety criteria

Example: P_1 Request (1,0,2)

- Check that Request ≤ Available (that is, $(1,0,2) \leq (3,3,2) \Rightarrow$ true

	Allocation	Need	Available
	A B C	A B C	A B C
P_0	0 1 0	7 4 3	2 3 0
P_1	3 0 2	0 2 0	
P_2	3 0 2	6 0 0	
P_3	2 1 1	0 1 1	
P_4	0 0 2	4 3 1	

- Executing safety algorithm shows that sequence $< P_1, P_3, P_4, P_0, P_2>$ satisfies safety requirement
- Can request for (3,3,0) by P_4 be granted?
- Can request for (0,2,0) by P_0 be granted?

3. Deadlock Detection

- Allow system to enter deadlock state
- Detection algorithm
- Recovery scheme

Single Instance of Each Resource Type

- Maintain **wait-for** graph
  - Nodes are processes
  - $P_i \rightarrow P_j$ if P_i is waiting for P_j
- Periodically invoke an algorithm that searches for a cycle in the graph. If there is a cycle, there exists a deadlock
- An algorithm to detect a cycle in a graph requires an order of $\mathbf{n^2}$ operations, where **n** is the number of vertices in the graph

Resource-Allocation Graph and Wait-for Graph

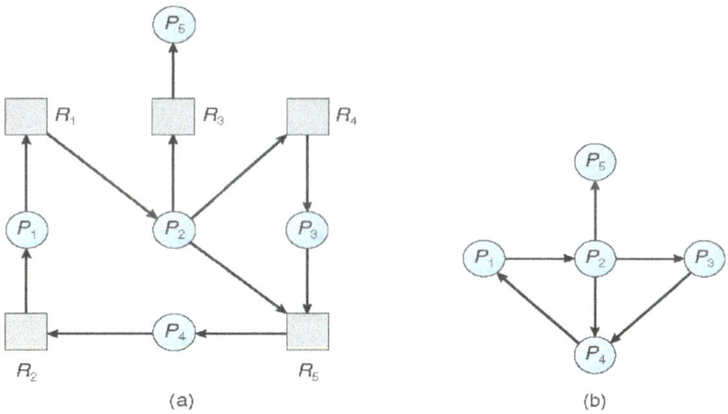

Figure 2.34 (a) Resource-allocation graph. (b) Corresponding wait-for graph

Several Instances of a Resource Type

- **Available**: A vector of length **m** indicates the number of available resources of each type
- **Allocation**: An n x m matrix defines the number of resources of each type currently allocated to each process

- **Request**: An n x m matrix indicates the current request of each process. If Request [i][j] = k, then process P_i is requesting k more instances of resource type R_j.

2.11.1.3 Detection Algorithm

1. Let **Work** and **Finish** be vectors of length **m** and **n**, respectively Initialize:
   a) (Work = Available)
   b) For i = 1,2, …, n, if Allocation ≠ 0, then Finish[i] = false; otherwise, Finish[i] = true
2. Find an index **i** such that both:
   a) Finish[i] == false
   b) $Request_i \leq Work$
   If no such i exists, go to step 4
3. Work = Work + Allocation Finish[i] = true go to step 2
4. If Finish[i] == false, for some i, $1 \leq i \leq n$, then the system is in deadlock state. Moreover, if Finish[i] == false, then P_i is deadlocked

Example of Detection Algorithm

- Five processes P_0 through P_4; three resource types A (7 instances), B (2 instances), and C (6 instances)
- Snapshot at time T_0:

	Allocation	Request	Available
	A B C	A B C	A B C
P_0	0 1 0	0 0 0	0 0 0
P_1	2 0 0	2 0 2	
P_2	3 0 3	0 0 0	
P_3	2 1 1	1 0 0	
P_4	0 0 2	0 0 2	

- Sequence <P_0, P_2, P_3, P_1, P_4> will result in **Finish[i] = true** for all **i**
- **P_2** requests an additional instance of type **C**

Request

	A B C
P_0	0 0 0
P_1	2 0 2
P_2	0 0 1
P_3	1 0 0
P_4	0 0 2

- State of system?
  - Can reclaim resources held by process P_0, but insufficient resources to fulfill other processes; requests
  - Deadlock exists, consisting of processes P_1, P_2, P_3, and P_4

Detection-Algorithm Usage

- When, and how often, to invoke depends on:
  - How often a deadlock is likely to occur?
  - How many processes will need to be rolled back?
    - one for each disjoint cycle
- If detection algorithm is invoked arbitrarily, there may be many cycles in the resource graph and so we would not be able to tell which of the many deadlocked processes "caused" the deadlock.

2.11.1.4 Recovery

i. Recovery using Process Termination

- Abort all deadlocked processes
- Abort one process at a time until the deadlock cycle is eliminated
- In which order should we choose to abort?
  1. Priority of the process
  2. How long process has computed, and how much longer to completion
  3. Resources the process has used
  4. Resources process needs to complete
  5. How many processes will need to be terminated
  6. Is process interactive or batch?

ii. Recovery using Resource Preemption

- **Selecting a victim** – minimize cost
- **Rollback** – return to some safe state, restart process for that state
- **Starvation** – same process may always be picked as victim, include number of rollback in cost factor

2

PROCESS MANAGEMENT

REVIEW QUESTIONS

PART-A

1. What is the difference between a program and a process?

- Program is passive entity stored on disk (executable file), process is active
- Program becomes process when executable file loaded into memory.

2. With the help of a state transition diagram, List out the various states of a process.

As a process executes, it changes state

1. **new**: The process is being created
2. **running**: Instructions are being executed
3. **waiting**: The process is waiting for some event to occur
4. **ready:** The process is waiting to be assigned to a processor
5. **terminated**: The process has finished execution

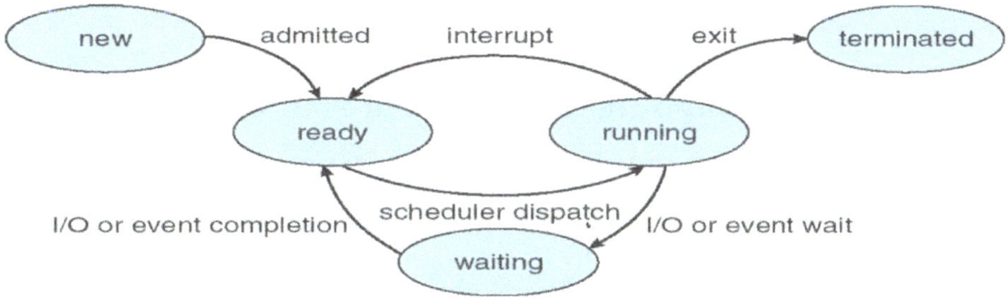

4. **What do you mean by PCB?**

Information associated with each process is called Process Control Block (PCB) also called task block, which includes

1. Process state
2. Program counter
3. CPU registers
4. CPU scheduling information-
5. Memory-management information
6. Accounting information
7. I/O status information

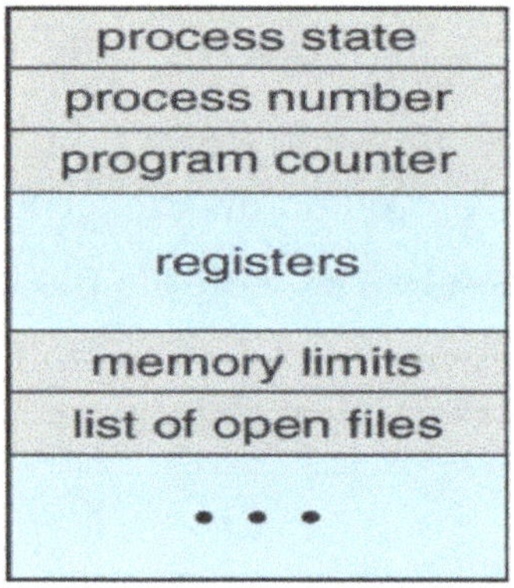

5. Give the scheduling queues of processes?

1. Job queue – set of all processes in the system

2. Ready queue – set of all processes residing in main memory, ready and waiting to execute

3. Device queues – set of processes waiting for an I/O device Processes migrate among the various queues

6. **What do you mean by queuing diagram representation?**

A common representation of process scheduling is a queueing diagram. Queueing diagram represents queues, resources, flows.

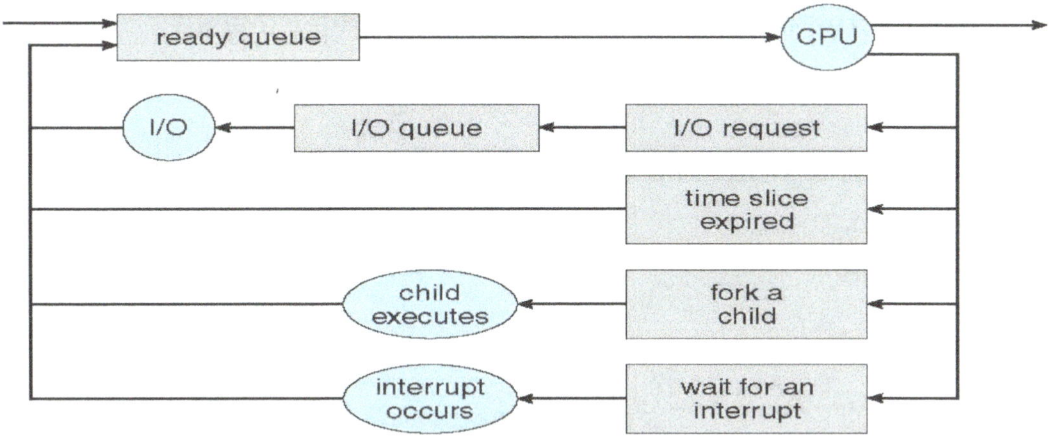

7. What is meant by schedulers? Give its types.

The operating system selects processes from queues for scheduling. The selection process is carried out by the appropriate scheduler.

Two types of schedulers

1. Short Term Schedulers (STS) or CPU Scheduler
2. Long Term Schedulers (LTS) or Job Scheduler.

8. Define STS.

- Short-term scheduler (or CPU scheduler) – selects which process should be executed next and allocates CPU
  - ➢ Sometimes the only scheduler in a system
  - ➢ Short-term scheduler is invoked frequently (milliseconds) ⇒ (must be fast).

9. Define LTS

- Long-term scheduler (or job scheduler) – selects which processes should be brought into the ready queue
  - ➢ Long-term scheduler is invoked infrequently (seconds, minutes) ⇒ (may be slow)
  - ➢ The long-term scheduler controls the degree of multiprogramming.

10. What is the work of Medium Term Scheduler (MTS)?

Medium-term scheduler can be added if degree of multiple programming needs to decrease

- Remove process from memory, store on disk, bring back in from disk to continue execution:
- **swapping**

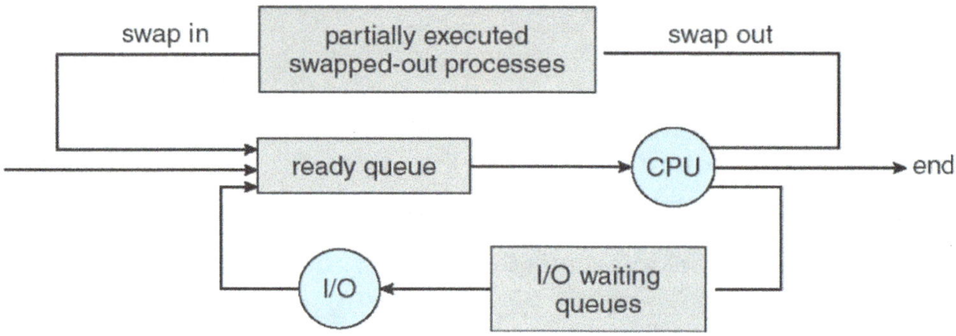

11. Define Context switch.

Switching the CPU to another process requires performing a state save of the current process and a state restore of a different process. This task is known as a context switch i.e., When CPU switches to another process, the system must save the state of the old process and load the saved state for the new process via a context switch

12. What is the motivation for

1. Multi-programming and
2. Time sharing?

The objective of multiprogramming is to have some process running at all times, to maximize CPU utilization. The objective of time sharing is to switch the CPU among so frequently that users can interact with each program while it is running.

- To Maximize CPU use, quickly switch processes onto CPU for time sharing
- Process scheduler selects among available processes for next execution on CPU

13. With the help of block diagrams, explain the flow of control between two processes during process switching?

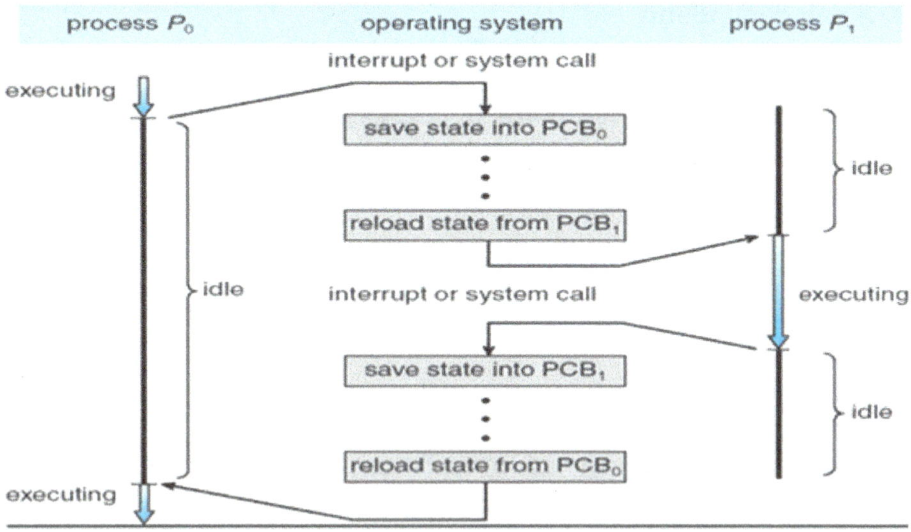

14. What happens when process context is switched? Is it an over-head?

Context of a process represented in the PCB

- Context-switch time is overhead; the system does no useful work while switching
  - ➢ The more complex the OS and the PCB ➔ the longer the context switch
- Context switch time dependent on hardware support
  - ➢ Some hardware provides multiple sets of registers per CPU ➔ multiple contexts loaded at once.

15. What are the operations can be done on processes?

1. Process creation
2. Process termination

16. What do you mean by cascading termination?

Some operating systems do not allow child to exists if its parent has terminated. If process terminates, then all its children must also be terminated. This is called as cascading termination. The termination is initiated by the operating system.

17. What is meant by independent processes ?

A process is independent if it cannot affect or be affected by the other processes executing in the system. Any process that does not share data with any other process is independent.

18. What is meant by cooperating processes?

- A process is cooperating if it can affect or be affected by the other processes executing in the system.
- Cooperating process can affect or be affected by other processes, including sharing data.

19. Why cooperating processes are needed in a system?

- **Information sharing** - an environment must be provided to allow concurrent access to information(shared file etc)
- **Computation speedup** - To improve the speedup, tasks are divided into subtasks, and executes in parallel
- **Modularity**- dividing the system functions into separate processes or threads
- **Convenience**- Even an individual user may work on many tasks at the same time.

20. What are the two models of communications?

1. Shared memory
2. Message passing

21. How data are communicated using shared memory?

- An area of memory shared among the processes that wish to communicate
- The communication is under the control of the users processes not the operating system.
- Major issues is to provide mechanism that will allow the user processes to synchronize their actions when they access shared memory.

22. How data are communicated in IPC using message passing?

- IPC facility provides two operations:
  - **send**(message)
  - **receive**(message)
- The message size is either fixed or variable

23. What is meant by blocking and non-blocking synchronization?

- Blocking is considered synchronous
  - **Blocking send** -- the sender is blocked until the message is received
  - **Blocking receive** -- the receiver is blocked until a message is available
- **Non-blocking** is considered **asynchronous**
  - **Non-blocking send** -- the sender sends the message and continue
  - **Non-blocking receive** -- the receiver receives:
    - A valid message, or
    - Null message.

24. Define thread.

A thread is a **basic unit of CPU utilization**;

- It comprises a **thread ID**, **a program counter**, **a register set**, and a **stack**.
- It shares with other threads belonging to the same process its code section, data section, and other operating-system resources, such as open files and signals.
- A traditional (or **heavyweight)** process has a single thread of control.

25. What are the benefits of threads?

1. Responsiveness
2. Resource Sharing
3. Economy
4. Scalability

26. Give the difference between parallelism and concurrency?

- **Parallelism** implies a system can perform more than one task simultaneously
- **Concurrency** supports more than one task making progress

27. List out the programming challenges of Multicore or multiprocessor systems?

- Dividing activities -
- Balance
- Data splitting
- Data dependency
- Testing and debugging

28. List out the types of parallelism

- Data parallelism
- Task parallelism

29. Define Amdhal's Law.

- Identifies performance gains from adding additional cores to an application that has both serial and parallel components
- S is serial portion
- N processing cores

$$speedup \leq \frac{1}{S + \frac{(1-S)}{N}}$$

30. What are the two types of threads?

1. User threads
2. Kernel threads

31. List out the threading models?

1. Many-to-One
2. One-to-One
3. Many-to-Many

32. Give the shortcomings of Many-to-many threading model?

- Developers can create as many user threads as necessary, and the corresponding kernel threads can run in parallel on a multiprocessor

- When **a thread performs a blocking system call**, the kernel can schedule another thread for execution

33. Draw the diagram for Windows process and its resources

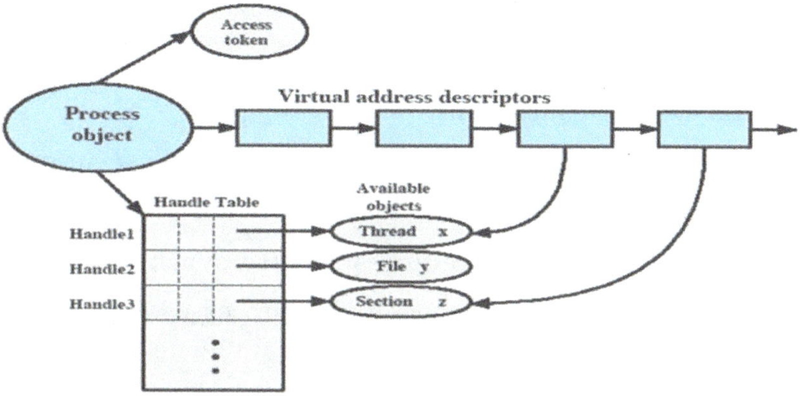

34. What is the use of tokens in windows processes?

- **The token** is used by windows to validate the user's ability to access secured objects or to perform restricted functions on the system and on secured objects.
- The **access token** controls whether the process can change its own attributes. If the process attempts to open such a handle, the **security system determines** whether this is permitted and therefore whether the process may change its own attributes.

35. List out the windows thread states?

1. Ready
2. Standby
4. Running
5. Waiting
6. Transition
7. Terminated

36. Define soft and hard affinity.

Soft Affinity: The dispatcher tries to assign a ready thread to the same processor it last ran on helps reuse data still in that processor's memory caches from the previous execution of the thread

Hard Affinity : An application restricts thread execution to certain processors.

37. Why we need process synchronization?

- Processes can execute concurrently
- May be interrupted at any time, partially completing execution
- Concurrent access to shared data may result in data inconsistency
- Maintaining data consistency requires process synchronization mechanisms to ensure the orderly execution of cooperating processes.

38. Define Race condition.

A situation in which several processes access and manipulate the same data concurrently and the outcome of the execution depends on the particular order in which the access takes place is called race condition.

39. What do you mean by critical section problem?

- Consider system of n processes {p0, p1, ... pn-1}
- Each process has critical section segment of code
- Process may be changing common variables, updating table, writing file, etc
- When one process in critical section, no other may be in its critical section
- Critical section problem is to design protocol to solve this Each process must ask permission to enter critical section in entry section, may follow critical section with exit section, then remainder section

General structure:

```
do {
        entry section
            critical section
        exit section
            remainder section
} while (true);
```

40. What are the requirement that must required for Critical section algorithms. **Or List out the requirements for the solution of CS problem?**

1. Mutual Exclusion
2. Progress
3. Bounded Waiting

41. What is meant by mutual exclusion?

-If process $\mathbf{P}_i$ is executing in its critical section, then no other processes can be executing in their critical sections

42. Give two hardware instructions and their definitions which can be used for implementing mutual exclusion

- Test and set
- Compare and swap

43. What is meant by semaphore? List out the operations used.

- Synchronization tool that provides more sophisticated ways (than Mutex locks) for process to synchronize their activities.
- Semaphore S – integer variable
- Can only be accessed via two indivisible (atomic) operations **wait() and signal()**
- Originally called P() and V()

43. Differentiate preemptive and non-preemptive approaches

Preemptive – allows preemption of process when running in kernel mode

Non-preemptive – runs until exits kernel mode, blocks, or voluntarily yields CPU Essentially free of race conditions in kernel mode

44. Give the types of semaphores?

1. Counting semaphore
2. Binary semaphore

45. Define busy waiting.

While a process is in its critical section, any other process that tries to enter its critical section must loop continuously in the while loop, which is also called a spin lock because the process "spins" while waiting for the lock to become available

46. How can be busy waiting avoided?

- With each semaphore there is an associated waiting queue
- Two operations:
  - **block** – place the process invoking the operation on the appropriate waiting queue
  - **wakeup** – remove one of processes in the waiting queue and place it in the ready queue

47. Define starvation or indefinite blocking

A process may never be removed from the semaphore queue in which it is suspended

48. What is meant by priority inversion?

Priority Inversion – Scheduling problem when lower-priority process holds a lock needed by higher-priority process

- Solved via **priority-inheritance protocol**

49. What is a mutex? What are the locks used and list out its drawbacks?

- Software tools to solve critical section problem
- Protect a critical section by first **acquire()** a lock then **release()** the lock

- Boolean variable indicating if lock is available or not

Drawbacks: spinlock, busy waiting

50. List out the Classical Problems of Synchronization.

Classical problems used to test newly-proposed synchronization schemes

1. Bounded-Buffer Problem
2. Readers and Writers Problem
3. Dining-Philosophers Problem

51. Declare the structure for monitors

```
{
        // shared variable declarations
        Function F1 (...) { .... }
        Function Fn (...) {......}
        Initialization code (...) { ... }
}
```

Two operations are allowed on a condition variable:

- **x.wait()** – a process that invokes the operation is suspended until **x.signal()**
- **x.signal()** – resumes one of processes (if any) that invoked **x.wait()**
  - If no **x.wait()** on the variable, then it has no effect on the variable

52. Define CPU scheduling.

1. **CPU scheduler or Short-term scheduler** selects from among the processes in ready queue, and allocates the CPU to one of them
   a. Queue may be ordered in various ways
2. CPU scheduling decisions may take place when a process:
   a. Switches from running to waiting state
   b. Switches from running to ready state
3. Switches from waiting to ready
4. Terminates

53. W hat is a Dispatcher?

- Dispatcher module gives control of the CPU to the process selected by the short-term scheduler; this involves:

- ➢ switching context
- ➢ switching to user mode
- ➢ jumping to the proper location in the user program to restart that program

54. W hat is dispatch latency?

Time it takes for the dispatcher to stop one process and start another running

55. W hat are the various scheduling criteria for CPU scheduling?

- **CPU utilization** – keep the CPU as busy as possible
- **Throughput** – # of processes that complete their execution per time unit
- **Turnaround time** – amount of time to execute a particular process
- **Waiting time** – amount of time a process has been waiting in the ready queue
- **Response time** – amount of time it takes from when a request was submitted until the first response is produced, not output (for time-sharing environment)

56. List out the various scheduling algorithms.

Scheduling Algorithms

1. First Come, First-Served Scheduling (FCFS)
2. Shortest-Job-First (SJF) Scheduling
3. Priority Scheduling
4. Round Robin (RR)
5. Multilevel Queue
6. Multilevel Feedback Queue

57. What is FCFS Scheduling?

In this algorithm, the process that requests the CPU first is allocated the CPU first. This implementation of the FCFS policy is easily managed with a FIFO queue.

58. What is drawback of FCFS?

Convoy effect - short process behind long process. i.e., Consider one CPU-bound and many I/O-bound processes

59. What is aging?

The problem of starvation (low priority processes may never execute) is solved using aging in which, as time progresses increase the priority of the process

60. Define deadlock.

In a multiprogramming environment, several processes may compete for a finite number of resources. A process requests resources; if the resources are not available at that time, the process enters a waiting state. Sometimes, a waiting process is never again able to change state, because the resources it has requested are held by other waiting processes. This situation is called a **deadlock.**

61. What are conditions under which a deadlock situation may arise?

1. **Mutual exclusion:** only one process at a time can use a resource
2. **Hold and wait:** a process holding at least one resource is waiting to acquire additional resources held by other processes
3. **No preemption:** a resource can be released only voluntarily by the process holding it, after that process has completed its task
4. **Circular wait:** there exists a set $\{P_0, P_1, \ldots, P_n\}$ of waiting processes such that P_0 is waiting for a resource that is held by P_1, P_1 is waiting for a resource that is held by P_2, …, P_{n-1} is waiting for a resource that is held by P_n, and P_n is waiting for a resource that is held by P_0.

62. What is a resource allocation graph (RAG)?

A set of vertices V and a set of edges E.

- V is partitioned into two types:
  - ➢ $P = \{P_1, P_2, \ldots, P_n\}$, the set consisting of all the processes in the system
  - ➢ $R = \{R_1, R_2, \ldots, R_m\}$, the set consisting of all resource types in the system
  - ➢ request edge – directed edge $P_i \rightarrow R_j$
- assignment edge – directed edge $R_j \rightarrow P_i$

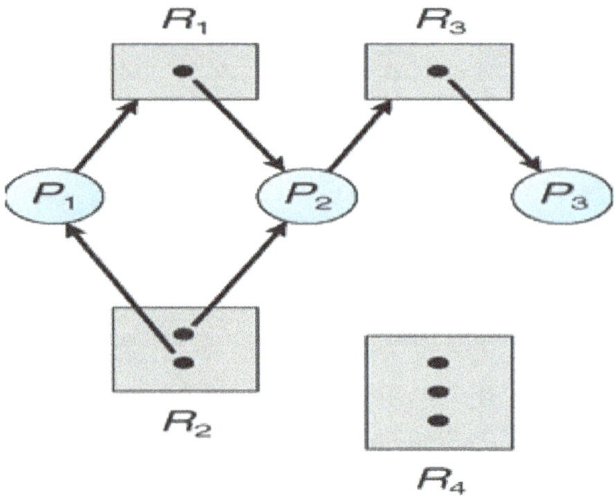

63. What are the methods for handling deadlocks?

1. **Deadlock prevention**
2. **Deadlock avoidance**
3. **Deadlock detection**
4. **Deadlock recovery**

64. Define deadlock prevention.

By denying any one of the necessary condition (4 conditions) for a deadlock to occur

65. Define deadlock avoidance.

Requires that the system has some additional **a priori** information available

- Simplest and most useful model requires that each process declare the **maximum number** of resources of each type that it may need
- The deadlock-avoidance algorithm dynamically examines the resource-allocation state to ensure that there can never be a circular-wait condition
- Resource-allocation state is defined by the number of available and allocated resources, and the maximum demands of the processes

66. What are a safe state and an unsafe state?

System is in **safe state** if there exists a sequence <P_1, P_2, …, P_n> of ALL the processes in the systems such that for each P_i, the resources that P_i can still request can be satisfied by currently available resources + resources held by all the P_j, with j < I

67. What is a claim edge?

- **Claim edge** $P_i \rightarrow R_j$ indicated that process P_j may request resource R_j; represented by a dashed line
- Claim edge converts to request edge when a process requests a resource
- Request edge converted to an assignment edge when the resource is allocated to the process
- When a resource is released by a process, assignment edge reconverts to a claim edge
- Resources must be claimed a priori in the system

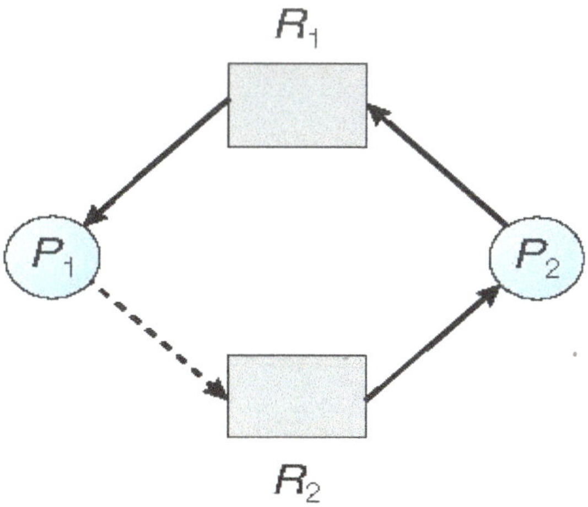

68. What is a WFG(Wait-For – Graph)?

WFG use to discover the presence of deadlocks in a system.

- Nodes are processes
- $\mathbf{P_i \rightarrow P_j}$ if $\mathbf{P_i}$ is waiting for $\mathbf{P_j}$

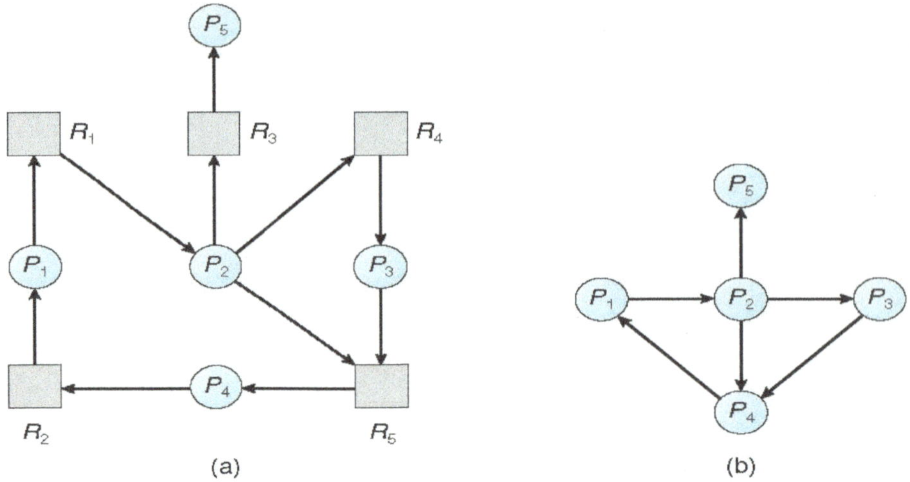

(a) Resource-allocation graph. (b) Corresponding wait-for graph.

69. How a system is recovered from a deadlock by resource preemption?

- **Selecting a victim** – minimize cost
- **Rollback** – return to some safe state, restart process for that state
- **Starvation** – same process may always be picked as victim, include number of rollback in cost factor

PART-B

1. **Write short notes on**

   a) Process Scheduling

   b) Queues

Ans: **a.** Definition - Process Scheduling

Types of Schedulers- STS, LTS, MTS

Queuing diagram

b. Types of queues with diagrams

2. **Write short notes on Operations on processes.**

   a. Process Creation

   b. Process termination

3. **Explain about IPC**

   a. Communication models

   b. Direct Vs Indirect

   c. Synchronization

   d. Buffering

4. **Write short notes on threading issues**

   a. Many to one

   b. One to one

   c. Many to many

   d. Two level

5. **Explain about Windows 7 thread and SMP Management**

   a. Diagram

   b. Process and thread object, attributes, services

   c. Thread states with diagram

   d. SMP Management- hard and soft affinity

6. **Explain about critical section problem**

   a. Definition

   b. General structure

   c. Requirements for the solutions
   d. Two process solutions
- e. Synchronization Hardware- test and set, compare and swap

7. Explain about semaphores

- a. Definition
- b. Two operations- wait, signal
- c. Implementations
- d. Problems

8. Explain about Monitors

- a. Definition
- b. Implementation

9. Explain about deadlock avoidance

Definition

- a. Single instance of a resource type
- b. Use a resource-allocation graph
- c. Multiple instances of a resource type
- d. Use the banker's algorithm
- e. Safety Algorithm
- f. Resource-Request Algorithm

10. Explain about deadlock prevention

How the necessary 4 conditions denied

11. Explain about deadlock detection

- a. Allow system to enter deadlock state
- b. Detection algorithm
- c. Recovery scheme

12. Explain about deadlock recovery

- a. Recovery from Deadlock: Process Termination
- b. Recovery from Deadlock: Resource Preemption

3

STORAGE MANAGEMENT

3.1 MAIN MEMORY

3.1.1 Background & Basic hardware

Memory is central to the operation of a modern computer system. Memory consists of a large array of bytes, each with its own address.

The CPU fetches instructions from memory according to the value of the program counter. These instructions may cause additional loading from and storing to specific memory addresses.

Basic Hardware

- All memory accesses are equivalent. ie.The memory hardware doesn't know what a particular part of memory is being used.
- The CPU can only access its registers and main memory. It cannot, for example, make direct access to the hard drive, so any data stored there must first be transferred into the main memory chips before the CPU can work with it.
- Memory accesses to registers are very fast, generally one clock tick, and a CPU may be able to execute more than one machine instruction per clock tick.
- Memory accesses to main memory are comparatively slow, and may take a number of clock ticks to complete. The basic idea of the cache is to transfer chunks of memory at a time from the main memory to the cache, and then to access individual memory locations one at a time from the cache.

- User processes must be restricted so that they only access memory locations that "belong" to that particular process. Every memory access made by a user process is checked against these two registers, base register and a limit register and if a memory access is attempted outside the valid range, then a fatal error is generated.

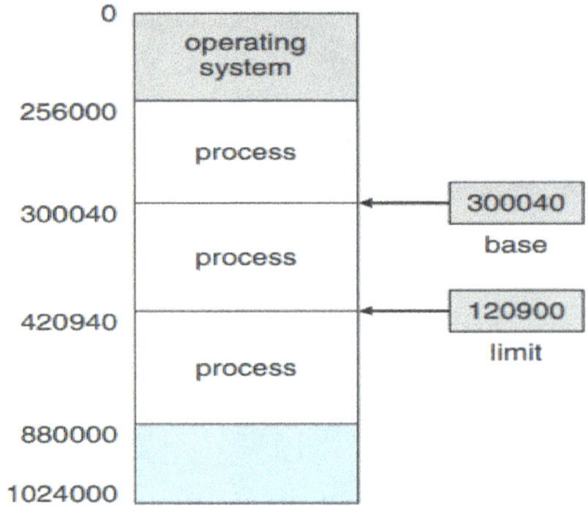

Figure 3.1 A base and a limit register define a logical address space

3.1.2 Hardware Address Protection

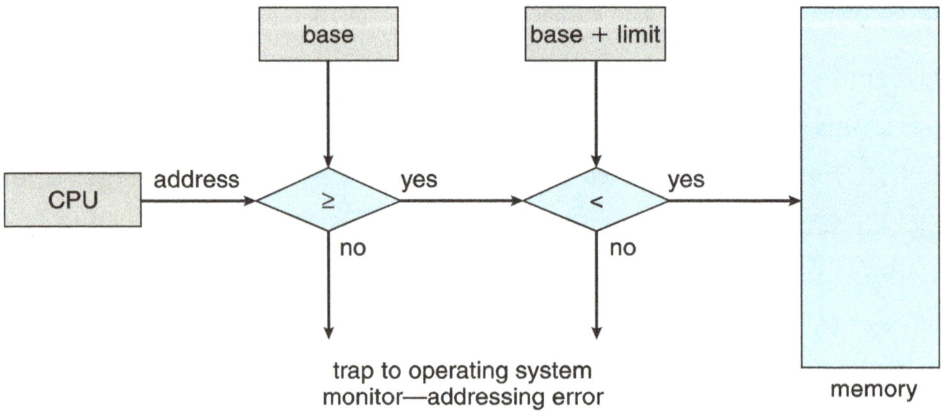

Figure 3.2 Hardware address protection with base and limit registers.

Protection of memory space is accomplished by having the CPU hardware compare every address generated in user mode with the registers. Any attempt by a program executing in user mode to access operating-system memory or other users' **memory results in a trap to the operating system**, which treats the attempt as a fatal error.

The base and limit registers can be loaded only by the operating system.

3.1.3 Address binding

Address binding of instructions and data to memory addresses can happen at **three different stages**

- **Compile time**: If memory location known a priori, absolute code can be generated; must recompile code if starting location changes.
- **Load time**: Must generate relocatable code if memory location is not known at compile time.
- **Execution time**: Binding delayed until run time if the process can be moved during its execution from one memory segment to another.

3.1.4 Logical Versus Physical Address Space

- The address generated by the CPU is a logical address, whereas the address actually seen by the memory hardware is a physical address.
- Addresses bound at compile time or load time have identical logical and physical addresses.
- Addresses created at execution time have different logical and physical addresses.
  - ➢ The logical address is also known as a virtual address.
  - ➢ The set of all logical addresses used by a program composes the logical address space, and the set of all corresponding physical addresses composes the physical address space.
- The run time mapping of logical to physical addresses is handled by the memory-management unit, MMU.
  - ➢ The MMU can take on many forms. One of the simplest is a modification of the base-register scheme.

  - ➢ The base register is now termed a relocation register, whose value is added to every memory request at the hardware level.
- The user programs never see physical addresses. User programs work entirely in logical address space, and any memory references or manipulations are done using purely logical addresses. Only when the address gets sent to the physical memory chips, is the physical memory address generated.

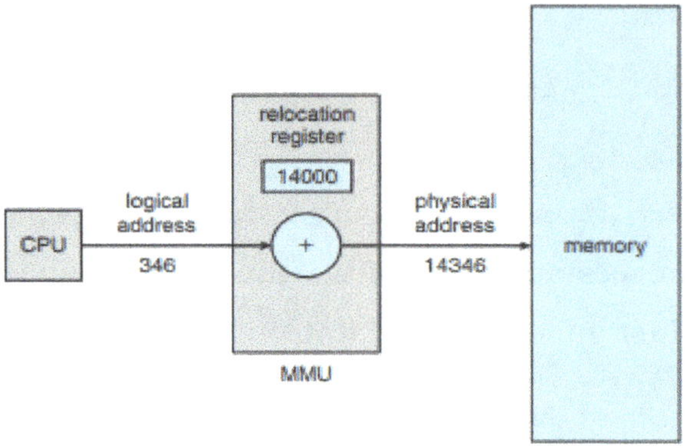

Figure 3.3 Dynamic relocation using relocation register

3.1.5 Dynamic Loading

- To obtain better memory-space utilization, dynamic loading is used.
- Here, a routine is not loaded until it is called. All routines are kept on disk in a relocatable load format.
- The main program is loaded into memory and is executed.
- When a routine needs to call another routine,
  - ➢ The calling routine first checks to see whether the other routine has been loaded.
  - ➢ Otherwise, the desired routine is loaded into memory and the programs address tables are updated.
  - ➢ Then control is passed to the newly loaded routine.

Advantage

- A routine is loaded only when it is needed. This method is particularly useful when large amounts of code are needed to handle infrequently occurring cases, such as error routines.
- Dynamic loading does not require special support from the OS. It is the responsibility of the users to design their programs

3.2 MEMORY MANAGEMENT STRATEGIES

There are various strategies that are used to manage memory using either of following two approaches.

- Contiguous memory allocation
- Non- contiguous memory allocation

3.2.1 CONTIGUOUS MEMORY ALLOCATION

Each process is allocated a single contiguous part of the memory. The different memory management schemes that are based on this approach are

- Single partition &
- Multiple partition

3.2.1.1 Single partition

- Main memory must support both operating system and user processes.The main memory need to be allocated efficiently.
- The memory is usually divided into two partitions:
    - one for the resident operating system and
    - one for the user processes.
- The operating system is in either low memory or high memory.

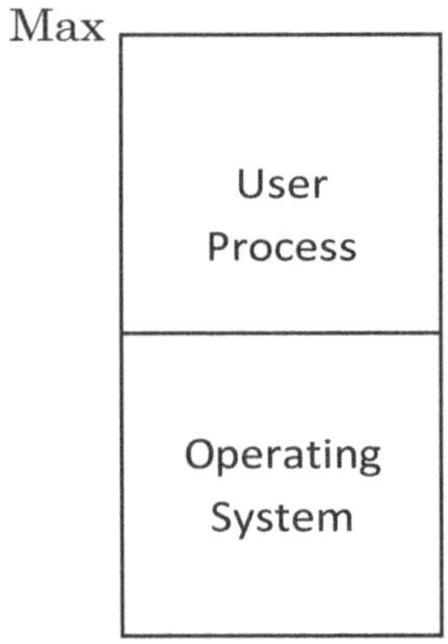

Figure 3.4 Memory Having Single Partition

In this scheme, only one process can execute at a time. Whenever a process is to be executed the OS loads it into the main memory for execution. After termination, OS waits for another process.

- When another arrives, it loads into the main memory overwriting the first one.
- Thus, OS needs to keep track of the first and the last location allocated to the user processes.

3.2.1.2 Multiple Partitions

A single partition scheme restricts the system to have only one process in memory at a time that reduces utilization of the CPU.

Simplest way to achieve multiprogramming is to divide memory into number of partitions which may be fixed or variable.

Fixed Partition

Each partition is of fixed size and can contain only one process. Thus, the degree of multiprogramming is bound by the number of partitions.

There are two types:

- Equal-sized partition &
- Unequal-sized partition

In **Equal-sized** partition, the process can be loaded into any partition.

Whenever a partition is free, a process is selected from the input queue and is loaded into the free partition.

When the process terminates, the partition becomes available for another process.

The OS is required to keep track only the partition occupied by each process.

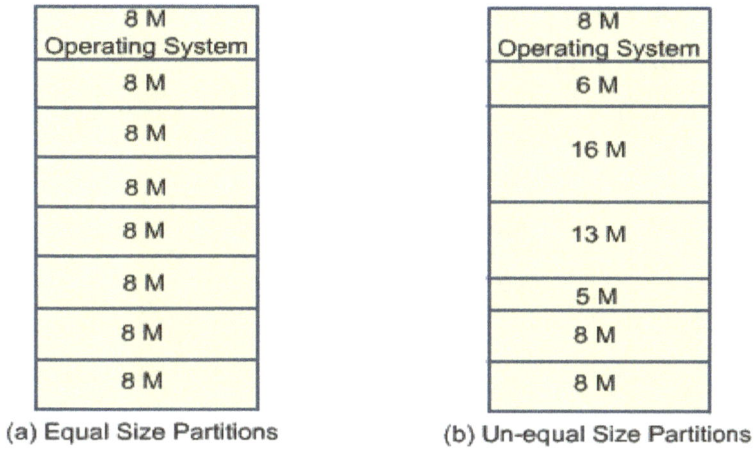

Figure 3.5 Memories Having Multiple Partitions

There is **one problem** with this method, ie. Memory utilization is not efficient.

Any process that occupies an entire partition which leads to the wastage of memory within the operation is called **internal fragmentation**.

This problem can be **resolved** using unequal-sized partition method.

In **Unequal-sized partition** method, a separate input queue is maintained for each partition.

Whenever a process arrives, it is placed into the input queue of the smallest process.

When this partition becomes free, it is allocated to the process.

Drawback of fixed-partitioning:

- The number of processes in memory depends on the number of partitions. Thus the degree of multiprogramming is limited.
- The memory can be used efficiently in case of processes of small sizes.

Variable-partition

- To overcome the disadvantages of fixed partition, a technique called MVT (Multiprogramming with a Variable number of Tasks.) is used.
- The partitions can vary in number and size.

- The amount of memory allocated is exactly the amount of memory a process requires.
- Initially, when there is no process in the memory, the whole memory is available for the allocation and is considered as a single large partition of available memory (a hole).
- Eventually, memory contains a set of holes of various sizes.

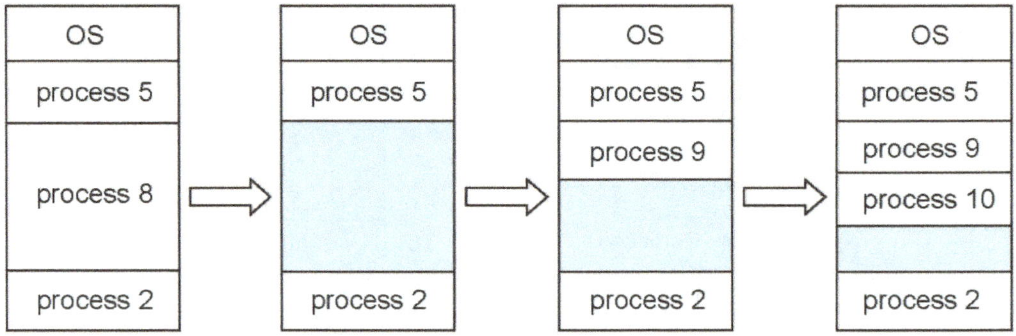

Figure 3.6 Multiple Partitions

- When a process request for the memory, the hole large enough to accommodate that process. The rest of the memory is available to other process.
- When a process terminates the memory occupied is deallocated.

External fragmentation exists when there is enough total memory space to satisfy a request but the available spaces are not contiguous: storage is fragmented into a large number of small holes.

Compaction: Compaction is a method to overcome the external fragmentation problem. All free blocks are brought together as one large block of free space. Compaction requires dynamic relocation. This scheme can be expensive.

3.2.2 Dynamic Storage Allocation Problem:

It concerns how to satisfy a request of size n from a list of free holes.

Solutions to this problem:

The first-fit, best-fit, and worst-fit strategies are the ones most commonly used to select a free hole from the set of available holes.

• **First fit**. Allocate the first hole that is big enough. Searching can start either at the beginning of the set of holes or at the location where the previous first-fit search ended. We can stop searching as soon as we find a free hole that is large enough.

• **Best fit**. Allocate the smallest hole that is big enough. We must search the entire list, unless the list is ordered by size. This strategy produces the smallest leftover hole.

• **Worst fit**. Allocate the largest hole. Again, we must search the entire list, unless it is sorted by size. This strategy produces the largest leftover hole, which may be more useful than the smaller leftover hole from a best-fit approach.

The first fit and best fit are better than worst fit in terms of decreasing time and storage utilization.

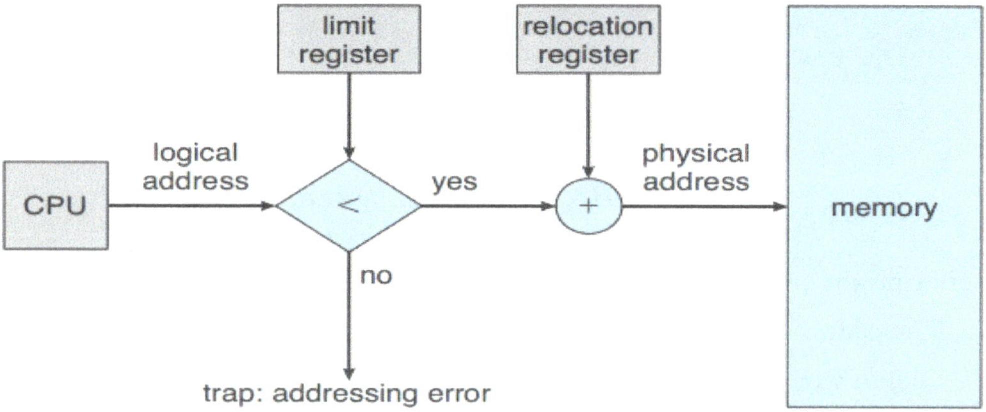

Figure 3.7 Hardware support for relocation and limit registers.

3.3 SEGMENTATION

- Segmentation is a memory-management scheme that supports this programmer view of memory. Natural extension of variable-sized partitions
  - ➢ Variable-sized partitions = 1 segment/process
  - ➢ Segmentation = many segments/process
- A logical address space is a collection of segments.
- Each segment has a name and a length.
- The programmer specifies each address by two quantities:

  Segment name and
  - ➢ Offset.

- Segments are numbered and are referred to by a segment number, rather than by a segment name. Thus, a logical address consists of a two tuple:

  <segment-number, offset>

- When a program is compiled, the compiler automatically constructs segments reflecting the input program.

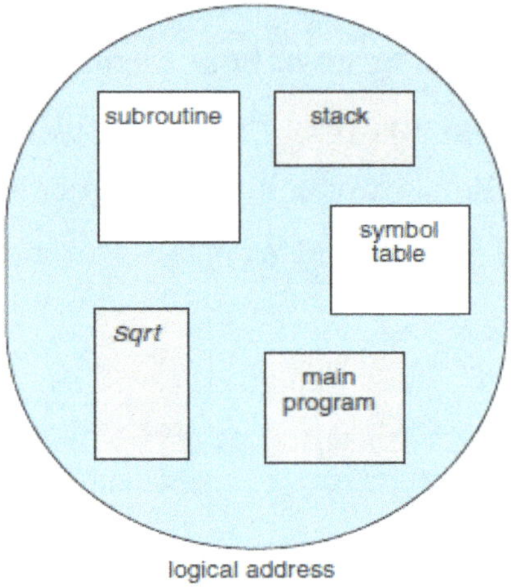

Figure 3.8 Logical space

A C compiler might create separate segments for the following:

1. The code
2. Global variables
3. The heap, from which memory is allocated
4. The stacks used by each thread
5. The standard C library

3.3.1 Segmentation Hardware

The mapping of two-dimensional user-defined addresses into one-dimensional physical addresses is affected by a segment table.

Each entry in the segment table has a

- **segment base** – specifies starting physical address where the segment resides in memory.
- **segment limit**- specifies the length of the segment.

A logical address consists of two parts:

- Segment number, s, and
- Offset into that segment, d.

The segment number is used as an index to the segment table.

The offset d of the logical address must be between 0 and the segment limit.

If it is not, it is trapped to the operating system (logical addressing attempt beyond end of segment).

When an offset is legal, it is added to the segment base to produce the address in physical memory of the desired byte.

The segment table is thus essentially an array of base–limit register pairs.

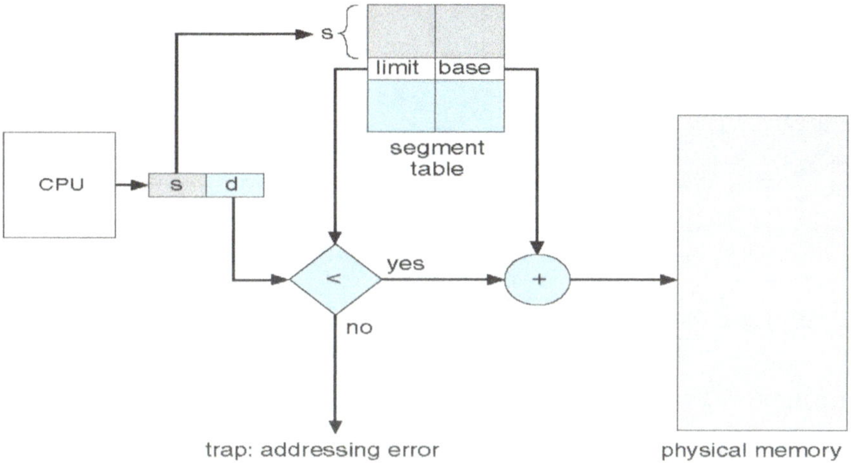

Figure 3.9 Segmentation Hardware

Example

Considering the example, there are five segments numbered from 0 through 4. The segments are stored in physical memory as shown. The segment table has a separate entry for each segment, giving the beginning address of the segment in physical memory (or base) and the length of that segment (or limit).

For example, segment 2 is 400 bytes long and begins at location 4300.

Thus, a reference to byte 53 of segment 2 is mapped onto location 4300 + 53 = 4353.

A reference to segment 3, byte 852, is mapped to 3200 (the base of segment 3) + 852 = 4052.

A reference to byte 1222 of segment 0 would result in a trap to the operating system, as this segment is only 1,000 bytes long.

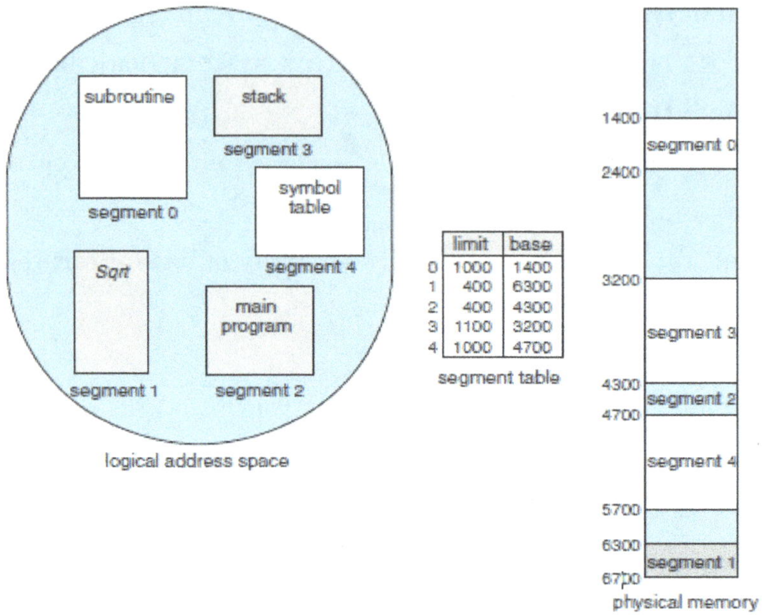

Figure 3.10 Example of segmentation

Various techniques

- Fixed partitions – easy to use, but internal fragmentation
- Variable partitions – more efficient, but external fragmentation
- Paging – use small, fixed size chunks, efficient for OS
- Segmentation – manage in chunks from user's perspective
- Combine paging and segmentation to get benefits

3.4 PAGING

- Paging is a memory management scheme that allows processes physical memory to be discontinuous, and which eliminates problems with fragmentation by allocating memory in equal sized blocks known as pages.
- Paging eliminates most of the problems and is the predominant memory management technique used today.

3.4.1 Basic Method- Paging Hardware

The **basic idea** behind paging is to

- Divide physical memory into a number of equal sized blocks called **frames**, and
- Divide programs logical memory space into blocks of the same size called **pages.**
- Any page (from any process) can be placed into any available frame.
- The page table is used to look up what frame a particular page is stored in at the moment.
- In the following example, for instance, page 2 of the program's logical memory is currently stored in frame 3 of physical memory:

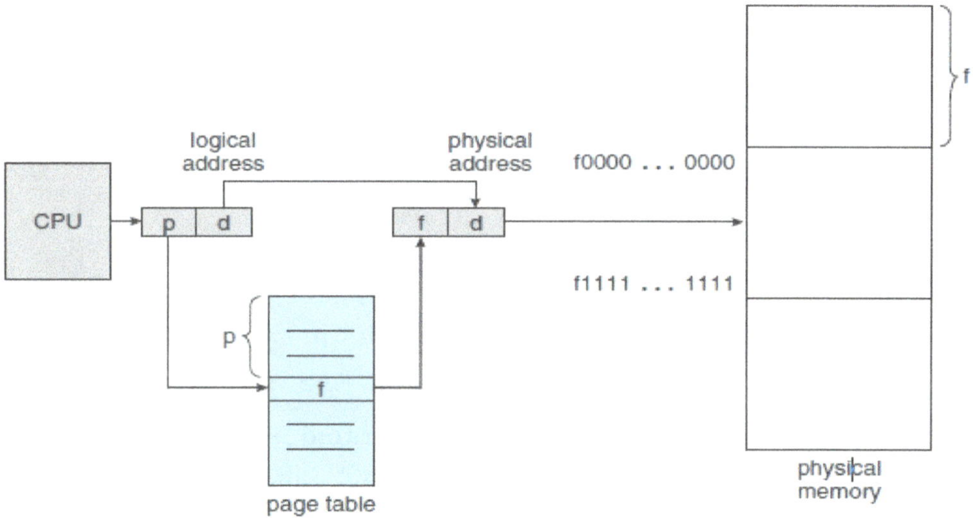

Figure 3.11 Paging hardware

- Every address generated by the CPU is divided into **two parts**:
  - ➢ **page number (p)**- in which the address resides. The number of bits in the page number limits how many pages a single process can address.
  - ➢ **page offset (d)**- from the beginning of that page. The number of bits in the offset determines the maximum size of each page, and should correspond to the system frame size.

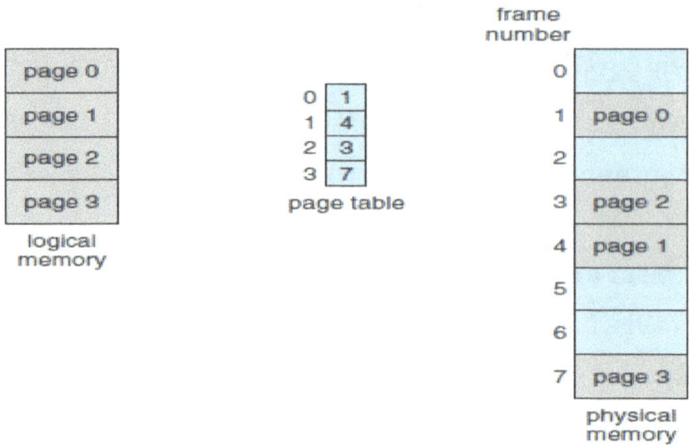

Figure 3.12 Paging model of logical and physical memory

- The page table maps the page number to a frame number, to yield a physical address which also has two parts:
  - The frame number and
  - The offset within that frame.
- The number of bits in the frame number determines how many frames the system can address, and the number of bits in the offset determines the size of each frame.
- Page numbers, frame numbers, and frame sizes are determined by the architecture, but are typically powers of two, allowing addresses to be split at a certain number of bits.

For example, if the logical address size is 2^m and the page size is 2^n, then the high-order m-n bits of a logical address designate the page number and the remaining n bits represent the offset.

The number of bits in the page number and the number of bits in the frame number do not have to be identical.

The former determines the address range of the logical address space, and the latter relates to the physical address space.

Thus, the logical address is as follows:

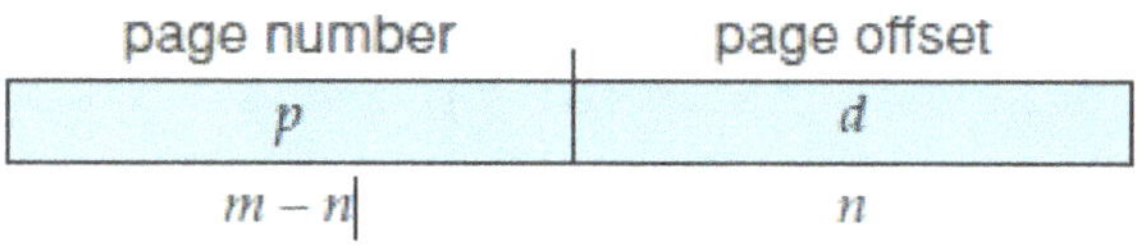

Where p is an index into the page table and

d is the displacement within the page.

3.4.2 Example

Consider the memory where in the logical address, n= 2 and m = 4. Using a page size of 4 bytes and a physical memory of 32 bytes (8 pages), show how the programmer's view of memory can be mapped into physical memory.

Solution:

Logical address 0 is page 0, offset 0. Indexing into the page table, it is found that page 0 is in frame 5.

Thus, logical address 0 maps to physical address 20 [= (5 × 4) +0].

Logical address 3 (page 0, offset 3) maps to physical address 23 [= (5 × 4) + 3].

Logical address 4 is page 1, offset 0;

According to the page table, page 1 is mapped to frame 6.

Thus, logical address 4 maps to physical address 24 [= (6 × 4) + 0].

Logical address 13 maps to physical address 9.

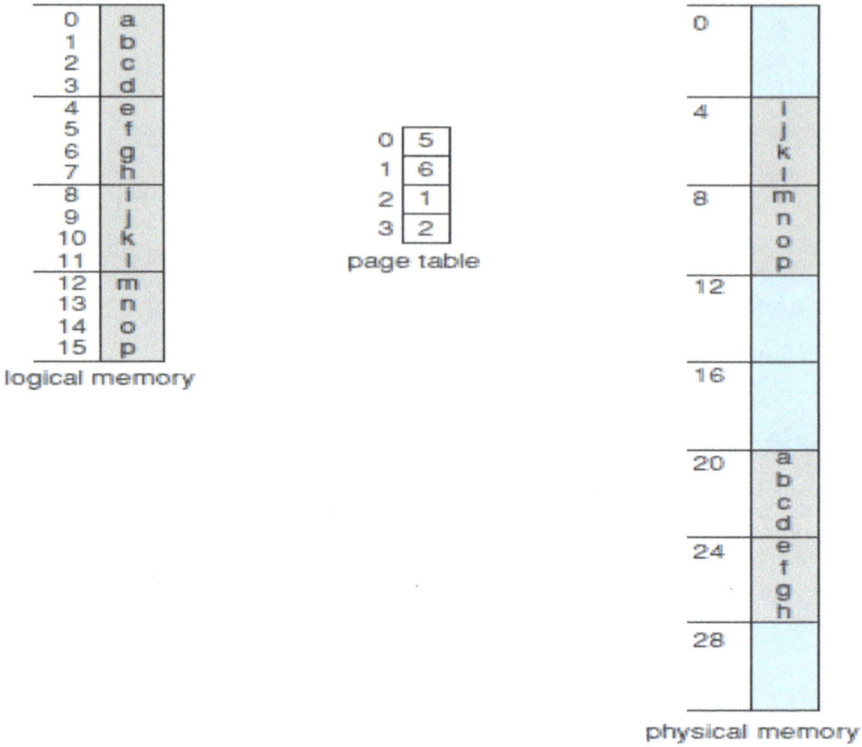

Figure 3.13 Paging example for a 32-byte memory with 4-byte pages

- The paging is like having a table of relocation registers, one for each page of the logical memory.
- There is **no external fragmentation** with paging. All blocks of physical memory are used, and there are no gaps in between and no problems with finding the right sized hole for a particular chunk of memory.
- But there is **internal fragmentation**. Memory is allocated in chunks the size of a page, and on the average, the last page will only be half full, wasting on the average half a page of memory per process.
- **Larger page sizes waste more memory**, but are **more efficient in terms of overhead**.
- Page table entries (frame numbers) are typically **32 bit numbers**, allowing access to 2^32 physical page frames.

- When a process requests memory (e.g. when its code is loaded in from disk), free frames are allocated from a free-frame list, and inserted into that process's page table.
- Processes are blocked from accessing anyone else's memory because all of their memory requests are mapped through their page table.
- The OS must keep track of each individual process's page table, updating it whenever the process's pages get moved in and out of memory, and applying the correct page table when processing system calls for a particular process.

This all increases the overhead involved when swapping processes in and out of the CPU. (The currently active page table must be updated to reflect the process that is currently running.)

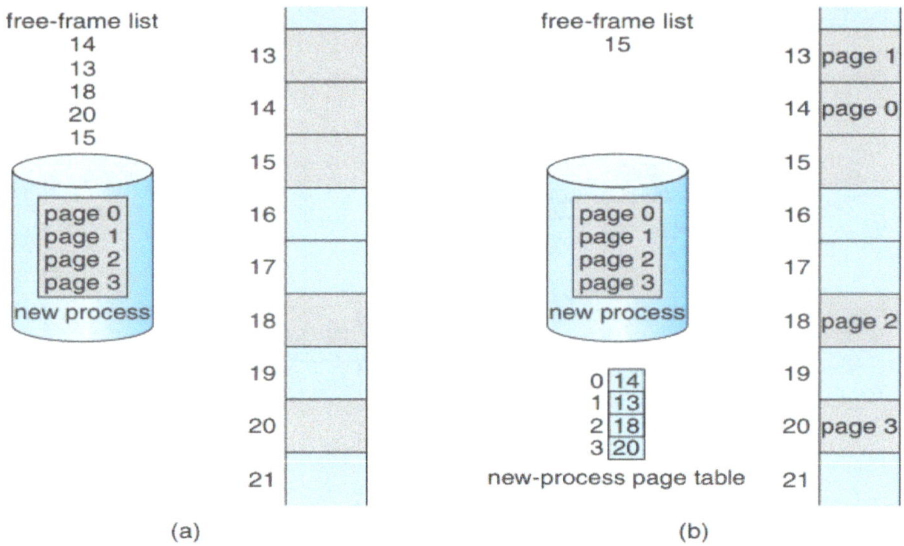

Figure 3.14 Free frames (a) before allocation and (b) after allocation

3.4.3 Hardware Support

- Page lookups must be done for every memory reference, and whenever a process gets swapped in or out of the CPU, its page table must be swapped in and out too, along with the instruction registers, etc. It is appropriate to provide hardware support for this operation.
- One option is to use a set of registers for the page table.

- An alternate option is to store the page table in main memory, and to use a single register (called the **page-table base register, PTBR**) to record where in memory the page table is located.
- Process switching is fast, because only the single register needs to be changed.
- However memory access just got half as fast, because every memory access now requires **two** memory accesses –
  - ➢ One to fetch the frame number from memory and
  - ➢ Another one to access the desired memory location.
- **Solution**: To use a very special high-speed memory device called the **translation look-aside buffer, TLB.**

Benefit of TLB: It can search an entire table for a key value in parallel, and if it is found anywhere in the table, then the corresponding lookup value is returned.

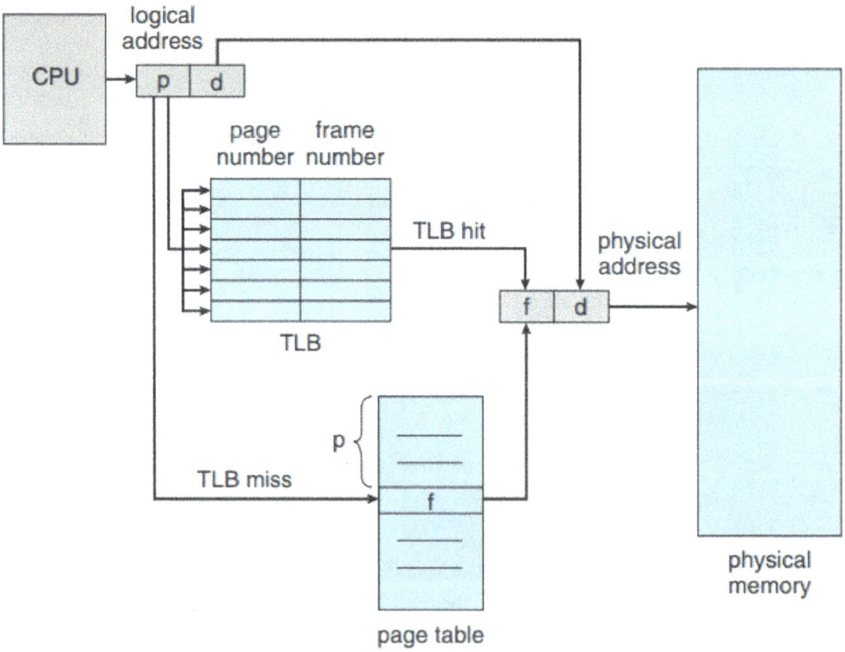

Figure 3.15 Paging hardware with TLB

- The TLB is **very expensive** and therefore **very small**. It is therefore **used as a cache device**.

- **Addresses are first checked** against the TLB, and if the info is not there (a TLB miss), then the frame is looked up from main memory and the TLB is updated.
- If the TLB is full, then replacement strategies range from **least-recently used, LRU** to random.
- Some TLBs allow some entries to be **wired down**, which means that they cannot be removed from the TLB. Typically these would be kernel frames.
- Some TLBs store **address-space identifiers, ASIDs**, to keep track of which process "owns" a particular entry in the TLB. Without this feature the TLB has to be flushed clean with every process switch.
- The percentage of time that the desired information is found in the TLB is termed the **hit ratio**.
- **For example**, suppose that it takes 100 nanoseconds to access main memory, and only 20 nanoseconds to search the TLB. So a TLB hit takes 120 nanoseconds total (20 to find the frame number and then another 100 to go get the data), it ignores the 20 nanoseconds and a TLB miss takes 220 (20 to search the TLB, 100 to go get the frame number, and then another 100 to go get the data.) again it ignores the 20 nanoseconds. So 99% hit rate would yield 101 nanoseconds average access time for a 1% slowdown.

$$0.80 * 100 + 0.20 * 200 = 120 \text{ nanoseconds}$$

3.4.4 Protection

- The page table can also help to protect processes from accessing memory.
- A bit or bits can be added to the page table to classify a page as read-write, read-only, read-write-execute. Then each memory reference can be checked to ensure it is accessing the memory in the appropriate mode.
- **Valid / invalid bits** can be added to "mask off" entries in the page table that are not in use by the current process.
- The valid / invalid bits described **cannot block all illegal memory accesses**, due to the internal fragmentation.
- Many processes do not use the entire page table available to them, particularly in modern systems with very large potential page tables. Rather than waste

memory by creating a full-size page table for every process, some systems use a **page-table length register, PTLR**, to specify the length of the page table.

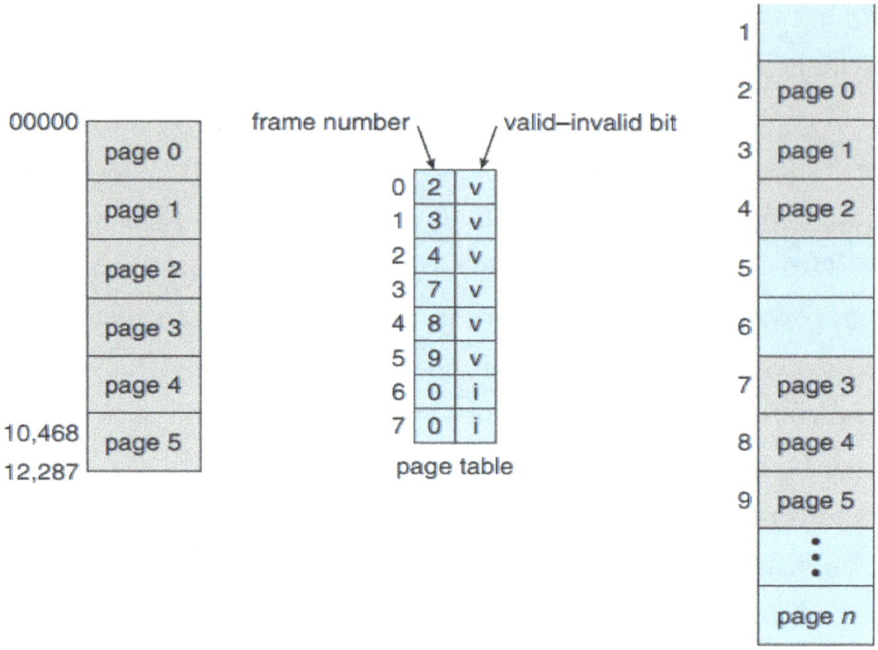

Figure 3.16 Valid (v) or invalid (i) bit in a page table

3.4.5 Shared Pages

- Paging systems can make it very easy to share blocks of memory, by simply duplicating page numbers in multiple page frames. This may be done with either code or data.
- If code is **reentrant**, that means that it does not write to or change the code in any way (it is non self-modifying), and it is therefore safe to re-enter it.(ie. the code can be shared by multiple processes).
- There are three different users running the editor simultaneously, but the code is only loaded into memory (in the page frames) one time.

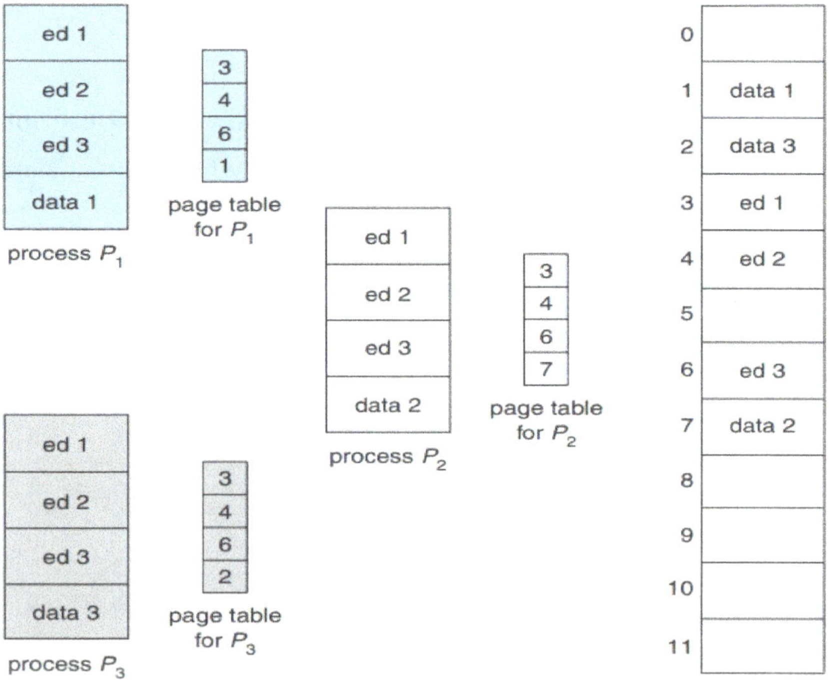

Figure 3.17 Sharing of code in a paging environment

3.4.6 INTEL 32 AND 64-BIT ARCHITECTURES

IA-32 Segmentation

- The Pentium CPU provides both pure segmentation and segmentation with paging. In the latter case, the CPU generates a logical address (segment-offset pair), which the segmentation unit converts into a logical linear address, which in turn is mapped to a physical frame by the paging unit.

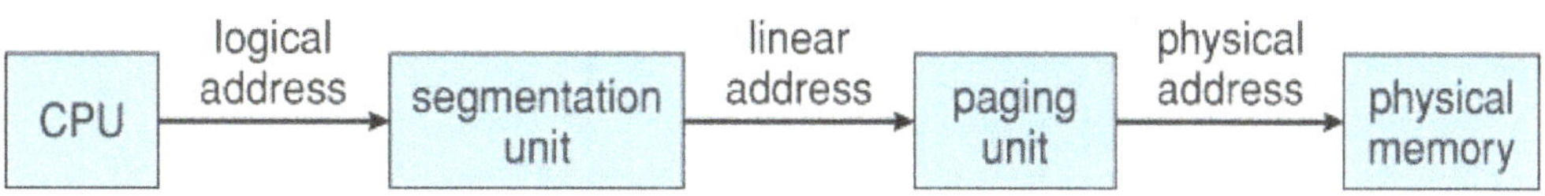

Figure 3.18 Logical to physical address translation in IA-32

- The Pentium architecture allows segments to be as large as 4 GB, (24 bits of offset).
- Processes can have as many as 16K segments, divided into two 8K groups:

- 8K private to that particular process, stored in the Local Descriptor Table, LDT.
- 8K shared among all processes, stored in the Global Descriptor Table, GDT.

- Logical addresses are (selector, offset) pairs, where the selector is made up of 16 bits:
  - A 13 bit segment number (up to 8K)
  - A 1 bit flag for LDT vs. GDT.
  - 2 bits for protection codes.

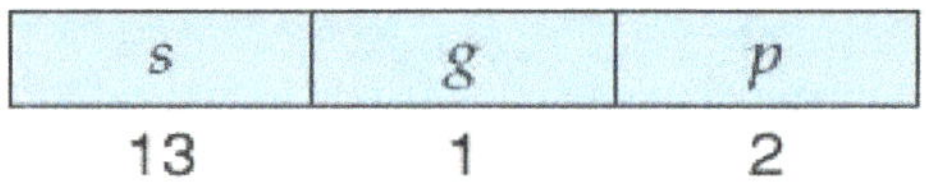

- The descriptor tables contain 8-byte descriptions of each segment, including base and limit registers.
- Logical linear addresses are generated by looking the selector up in the descriptor table and adding the appropriate base address to the offset.

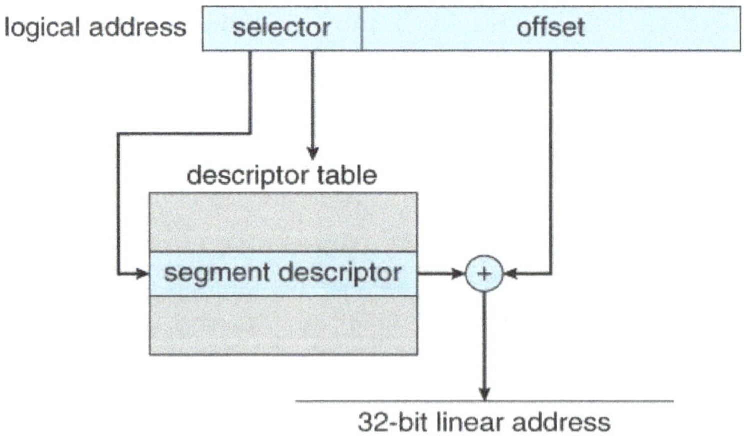

Figure 3.19 IA-32 segmentation

IA-32 Paging

- Pentium paging normally uses a two-tier paging scheme, with the first 10 bits being a page number for an outer page table (a.k.a. page directory), and the next 10 bits being a page number within one of the 1024 inner page tables, leaving the remaining 12 bits as an offset into a 4K page.

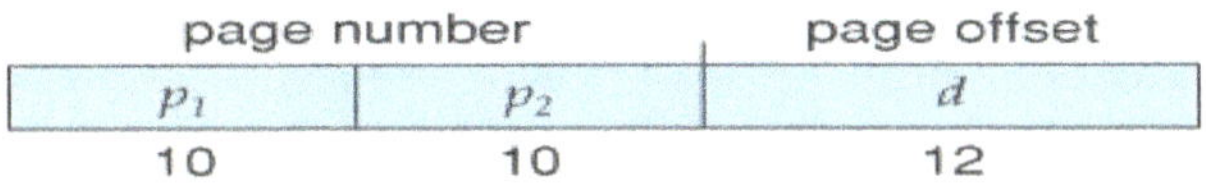

- A special bit in the page directory can indicate that this page is a 4MB page, in which case the remaining 22 bits are all used as offset and the inner tier of page tables is not used.
- The CR3 register points to the page directory for the current process.
- If the inner page table is currently swapped out to disk, then the page directory will have an "invalid bit" set, and the remaining 31 bits provide information on where to find the swapped out page table on the disk

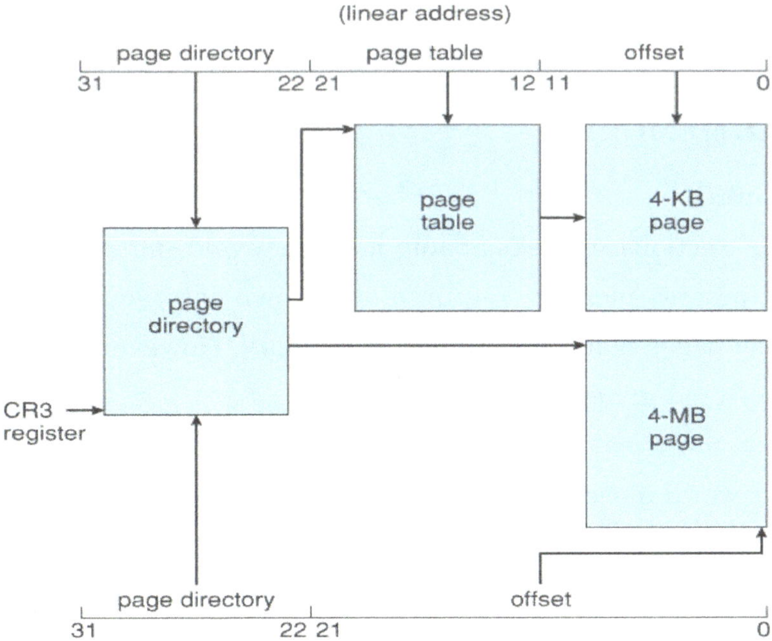

Figure 3.20 Paging in the IA-32 architecture.

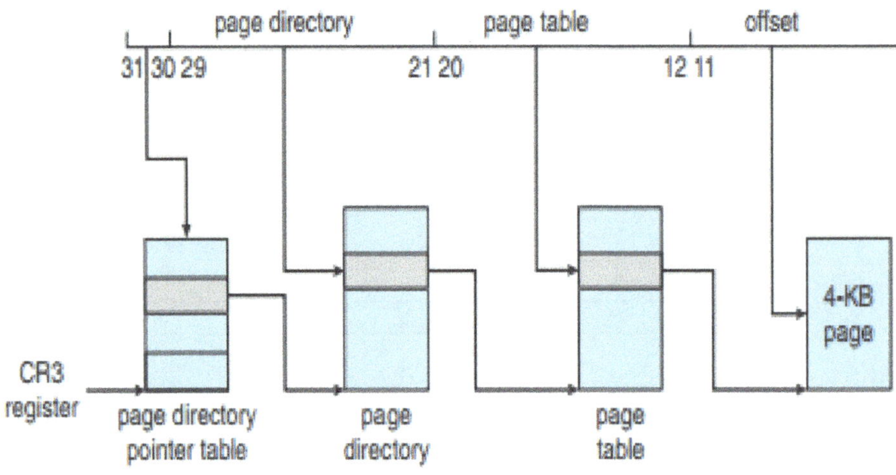

Figure 3.21 Page address extensions 8.7.2 x86-64

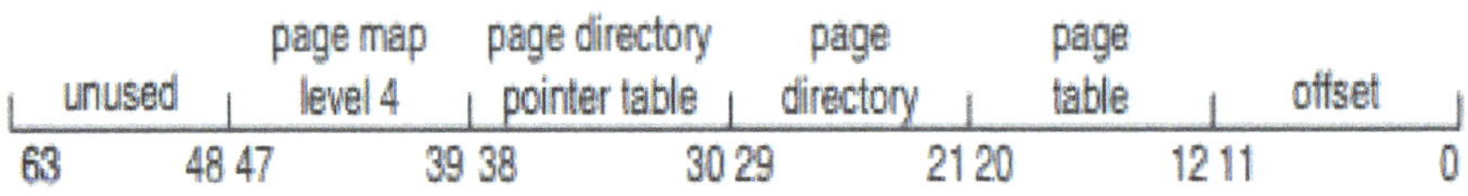

Figure 3.22 x86-64 linear address

3.5 VIRTUAL MEMORY

Background

- Preceding sections talked about how to avoid memory fragmentation by breaking process memory requirements down into smaller bites (pages), and storing the pages non-contiguously in memory. However the entire process still had to be stored in memory somewhere.
- In practice, most real processes do not need all their pages, or at least not all at once, for several reasons:
    1. Error handling code is not needed unless that specific error occurs, some of which are quite rare.
    2. Arrays are often over-sized for worst-case scenarios, and only a small fraction of the arrays are actually used in practice.

3. Certain features of certain programs are rarely used, such as the routine to balance the federal budget.

- The ability to load only the portions of processes that were actually needed has several benefits:
  - Programs could be written for a much larger address space (virtual memory space) than physically exists on the computer.
  - Because each process is only using a fraction of their total address space, there is more memory left for other programs, improving CPU utilization and system throughput.
  - Less I/O is needed for swapping processes in and out of RAM, speeding things up.

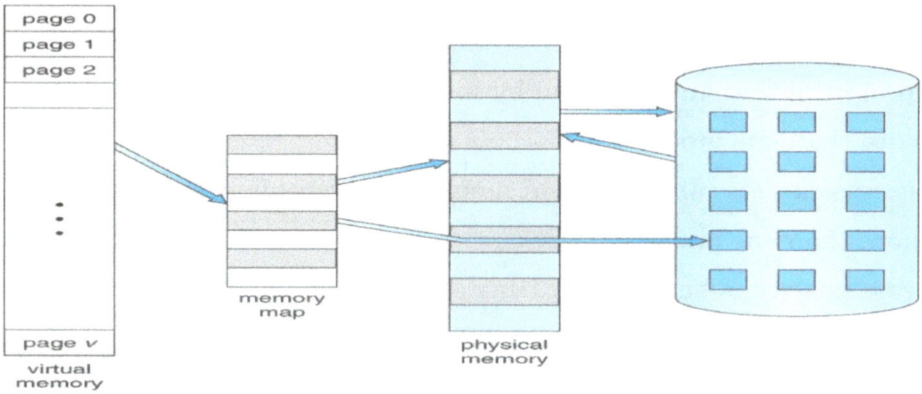

Figure 3.23 Diagram showing virtual memory that is larger than physical memory

- Consider the **virtual address space**, which is the programmer's logical view of process memory storage. The actual physical layout is controlled by the process's page table.
- The address space is **sparse** - A great hole in the middle of the address space is never used, unless the stack and/or the heap grow to fill the hole.

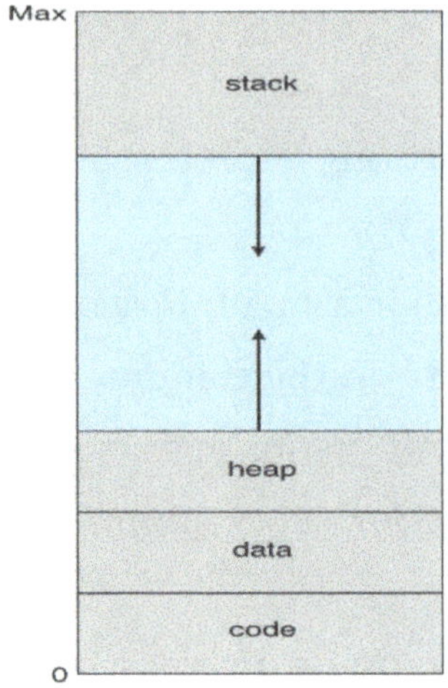

Virtual address space

- Virtual memory also allows the sharing of files and memory by multiple processes, with several benefits:
  - ➢ System libraries can be shared by mapping them into the virtual address space of more than one process.
  - ➢ Processes can also share virtual memory by mapping the same block of memory to more than one process.
  - ➢ Process pages can be shared during a fork( ) system call, eliminating the need to copy all of the pages of the original (parent) process.

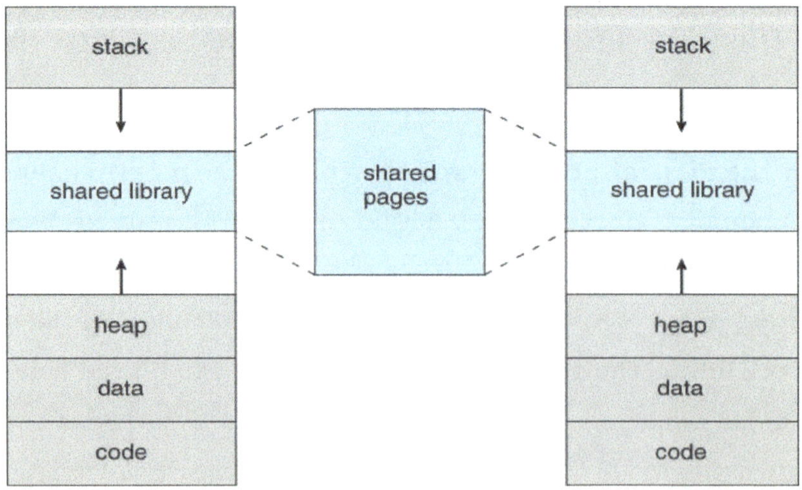

Figure 3.24 Shared library using virtual memory

3.6 DEMAND PAGING

- **Idea**- when a process is swapped in, its pages are not swapped in all at once. Rather they are swapped in only when the process needs them. (On demand.) This is termed a **lazy swapper**, although a **pager** is a more accurate term.

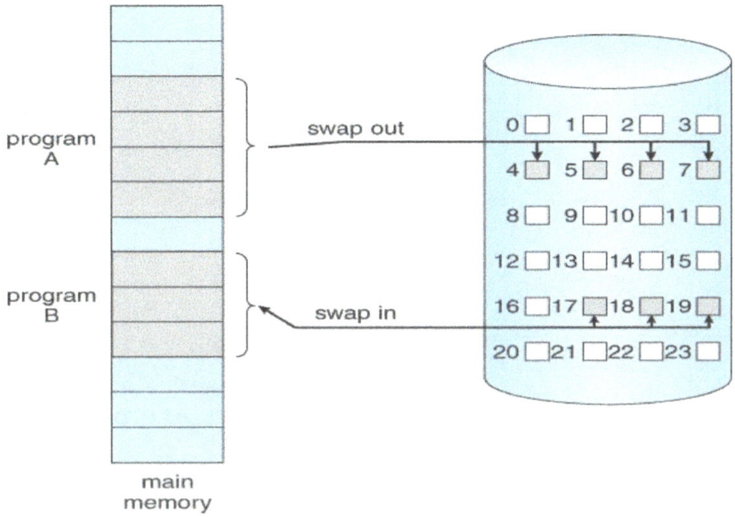

Figure 3.25 Transfer of a paged memory to contiguous disk space

3.6.1 Basic Concepts

- Idea behind paging- when a process is swapped in, the pager only loads into memory those pages that it expects the process to need.
- Pages that are not loaded into memory are marked as invalid in the page table, using the invalid bit. (The rest of the page table entry may either be blank or contain information about where to find the swapped-out page on the hard drive.)
- If the process only ever accesses pages that are loaded in memory (**memory resident** pages), then the process runs exactly as if all the pages were loaded in to memory.

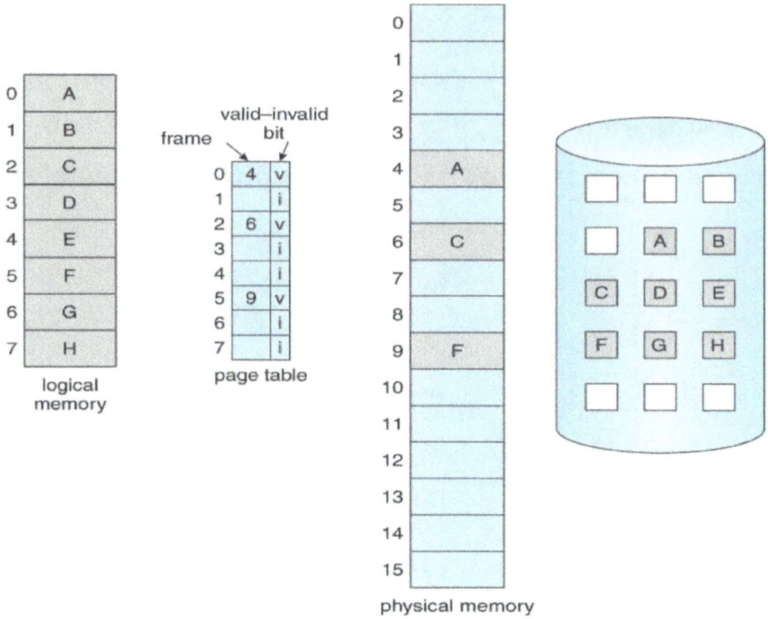

Figure 3.26 Page table when some pages are not in main memory.

- If a page is needed that was not originally loaded up, then a **page fault trap** is generated, which must be handled in a **series of steps**:
  1. The memory address requested is first checked, to make sure it was a valid memory request.
  2. If the reference was invalid, the process is terminated. Otherwise, the page must be paged in.
  3. A free frame is located, possibly from a free-frame list.
  4. A disk operation is scheduled to bring in the necessary page from disk.
  5. When the I/O operation is complete, the process's page table is updated with the new frame number, and the invalid bit is changed to indicate that this is now a valid page reference.
  6. The instruction that caused the page fault must now be restarted from the beginning.

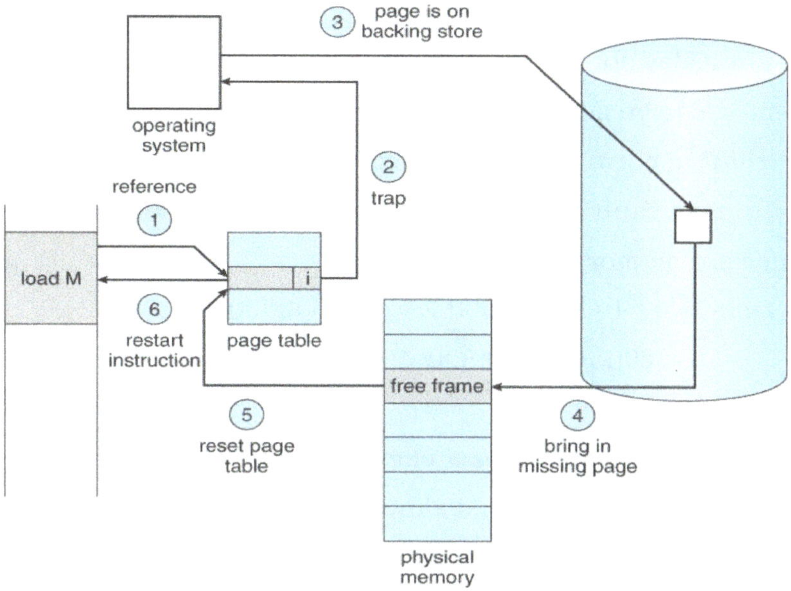

Figure 3.27 Steps in handling a page fault

- Here NO pages are swapped in for a process until they are requested by page faults. This is known as **pure demand paging.**
- Each instruction could generate multiple page fault & is very rare, due to **locality of reference.**
- The hardware necessary to support virtual memory is the same as for paging and swapping: a page table and secondary memory.
- A crucial part of the process is that the instruction must be restarted from scratch once the desired page has been made available in memory. For most simple instructions this is not a major difficulty.
- If some of the data gets modified before the page fault occurs, this could cause problems. One solution is to access both ends of the block before executing the instruction, guaranteeing that the necessary pages get paged in before the instruction begins.

3.6.2 Performance of Demand Paging

- There is some slowdown and performance hit whenever a page fault occurs and the system has to go get it from memory.
- Suppose that a normal memory access requires 200 nanoseconds, and that servicing a page fault takes 8 milliseconds. (8,000,000 nanoseconds, or 40,000 times a normal memory access.)

Effective access time is = (1 - p) * (200) + p * 8000000

= 200 + 7,999,800 * p, where p- **page fault rate** (on a scale from 0 to 1),

Even if only one access in 1000 causes a page fault, the effective access time drops from 200 nanoseconds to 8.2 microseconds, a slowdown of a factor of 40 times.

In order to keep the slowdown less than 10%, the page fault rate must be less than 0.0000025, or one in 399,990 accesses.

- In some systems, swap space is faster to access than the regular file system, because it does not have to go through the whole directory structure. So it transfers an entire process from the file system to swap space before starting up the process.
- Some systems use demand paging directly from the file system for binary code. This approach is used by both Solaris and BSD Unix.

Copy-on-Write

- Idea- the pages for a parent process do not have to be actually copied for the child until one or the other of the processes changes the page. They can be simply shared between the two processes.

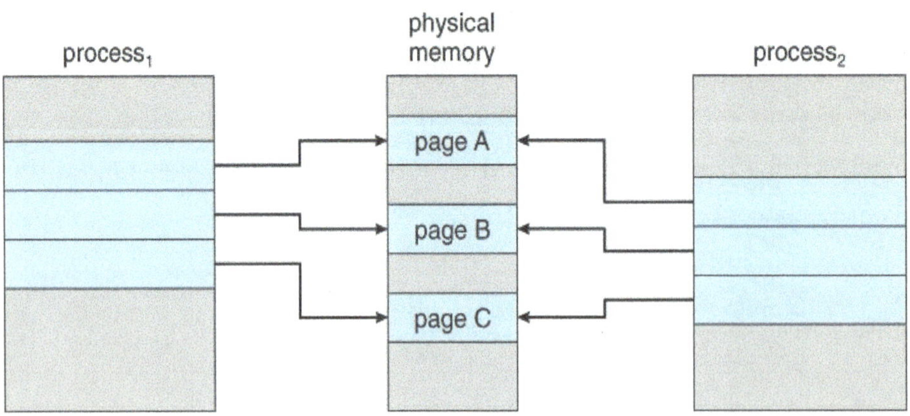

Figure 3.28 before **process 1 modifies page C.**

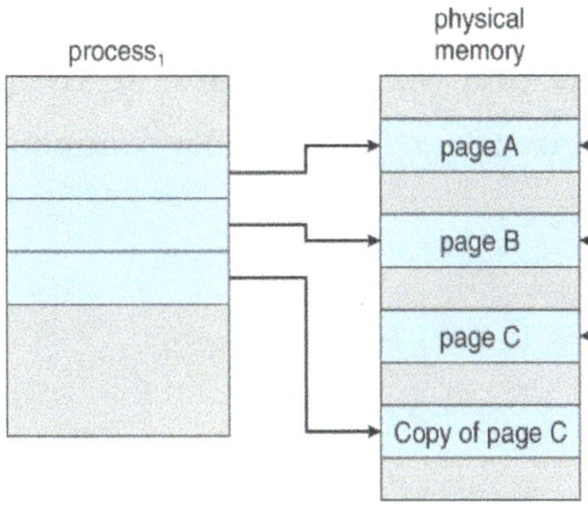

Figure 3.29 After process 1 modifies page C.

- Obviously only pages that can be modified even need to be labeled as copy-on-write. Code segments can simply be shared.
- Pages used to satisfy copy-on-write duplications are typically allocated using **zero-fill-on-demand**, meaning that their previous contents are zeroed out before the copy proceeds.
- Some systems provide an alternative to the fork( ) system call called a **virtual memory fork, vfork( )**.
- In this case the parent is suspended, and the child uses the parent's memory pages. This is very fast for process creation, but requires that the child not modify any of the shared memory pages before performing the exec( ) system call.

3.7 Page Replacement

In order to make use of virtual memory, several processes are loaded into memory at the same time. Since it is necessary to only load the pages that are actually needed by each process at any given time, there is room to load many more processes than to load in the entire process.

However memory is also needed for other purposes (such as I/O buffering), and if some process suddenly needs more pages and there aren't any free frames available.

Solutions:

1. Adjust the memory used by I/O buffering, etc., to free up some frames for user processes.
2. Swap some process out of memory completely, freeing up its page frames.

3. Find some page in memory that isn't being used right now, and swap that page only out to disk, freeing up a frame that can be allocated to the process requesting it. This is known as **page replacement**, and is the most common solution.

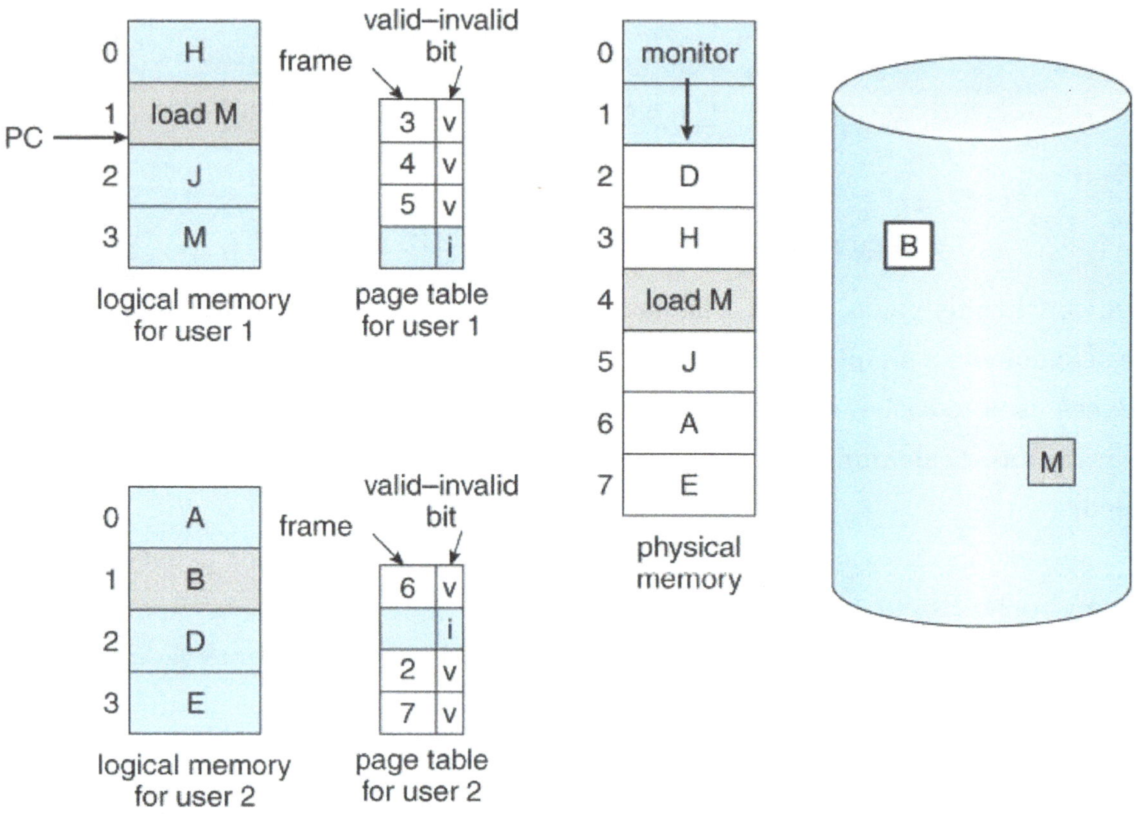

Figure 3.30 Need for page replacement.

3.7.1 Basics on Page Replacement

The page-fault processing assumed that there would be free frames available on the free-frame list. Now the page-fault handling must be modified to free up a frame if necessary, as follows:

1. Find the location of the desired page on the disk, either in swap space or in the file system.
2. Find a free frame:
   a. If there is a free frame, use it.

b. If there is no free frame, use a page-replacement algorithm to select an existing frame to be replaced, known as the **victim frame**.

c. Write the victim frame to disk. Change all related page tables to indicate that this page is no longer in memory.

3. Read in the desired page and store it in the frame. Adjust all related page and frame tables to indicate the change.
4. Restart the process that was waiting for this page.

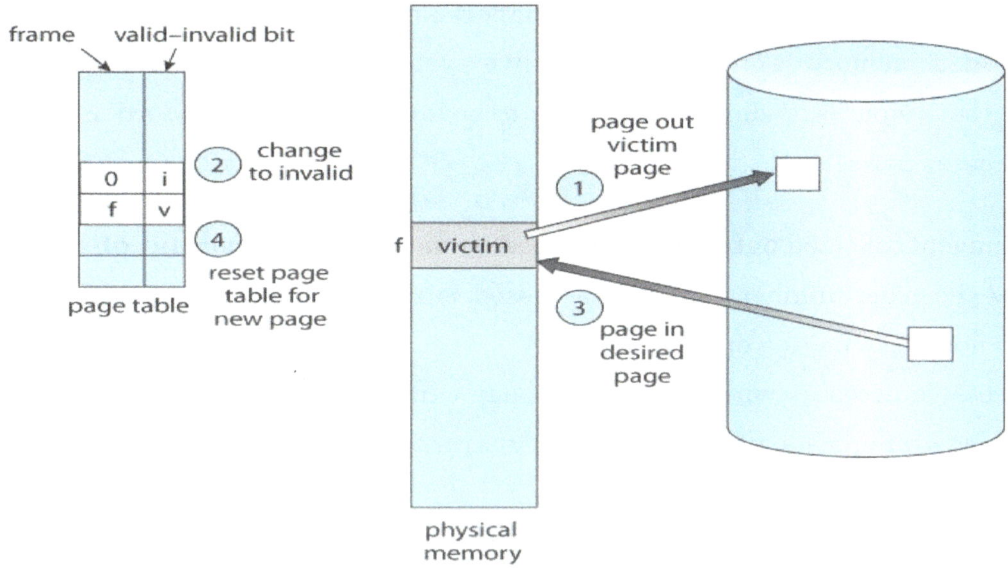

Figure 3.31 Page replacement.

There is an extra disk written to the page-fault handling, effectively doubling the time required to process a page fault.

This can be alleviated by assigning a **modify bit,** or **dirty bit** to each page, indicating whether or not it has been changed since it was last loaded in from disk.

If the dirty bit has not been set, then the page is unchanged, and does not need to be written out to disk.

Otherwise the page write is required.

Requirements to implement a demand paging system.

- **Frame-allocation algorithm-** how many frames are allocated to each process.
- **Page-replacement algorithm-** how to select a page for replacement when there are no free frames available.

Goal - to generate the fewest number of overall page faults. Because disk access is so slow relative to memory access, even slight improvements to these algorithms can yield large improvements in overall system performance.

Algorithms are evaluated using a given string of memory accesses known as a **reference string,** which can be generated in one of three common ways:

1. Randomly generated, either evenly distributed or with some distribution curve based on observed system behavior. This is the fastest and easiest approach.
2. Specifically designed sequences. These are useful for illustrating the properties of comparative algorithms in published papers and textbooks.
3. Recorded memory references from a live system. This may be the best approach, but the amount of data collected can be enormous, on the order of a million addresses per second.

The volume of collected data can be reduced by making two important **observations:**

1. Only the page number that was accessed is relevant. The offset within that page does not affect paging operations.
2. Successive accesses within the same page can be treated as a single page request, because all requests after the first are guaranteed to be page hits.

For example, if pages were of size 100 bytes, then the sequence of address requests (0100, 0432, 0101, 0612, 0634, 0688, 0132, 0038, 0420) would reduce to page requests (1, 4, 1, 6, 1, 0, 4). As the number of available frames increases, the number of page faults should decrease.

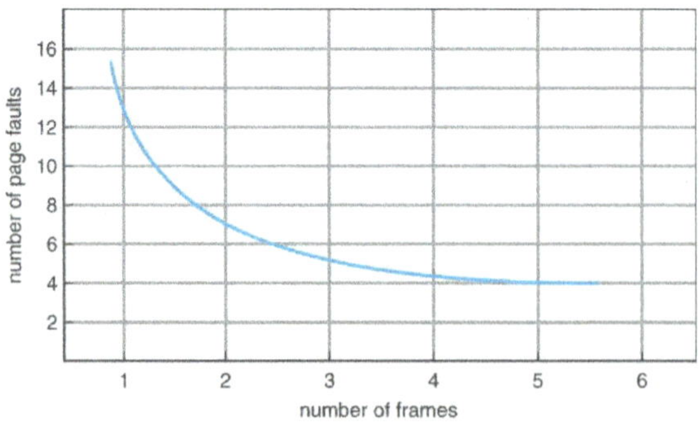

Figure 3.32 Graph of page faults versus number of frames.

3.7.2 Page Replacement Algorithms

3.7.2.1 FIFO Page Replacement

A simple and obvious page replacement strategy is **FIFO**, i.e. first-in-first-out. As new pages are brought in, they are added to the tail of a queue, and the page at the head of the queue is the next victim. In the following example, 20 page requests result in 15 page faults:

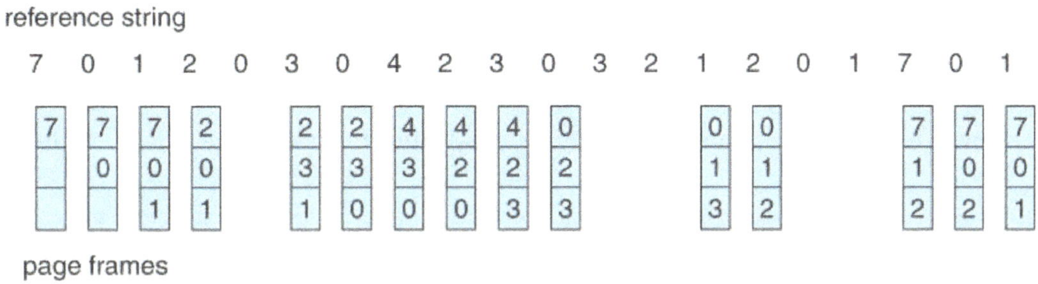

Figure 3.33 FIFO page-replacement algorithms.

- Although FIFO is simple and easy, it is not always optimal, or even efficient.
- An interesting effect that can occur with FIFO is **Belady's anomaly**, in which increasing the number of frames available can actually **increase** the number of page faults that occur!

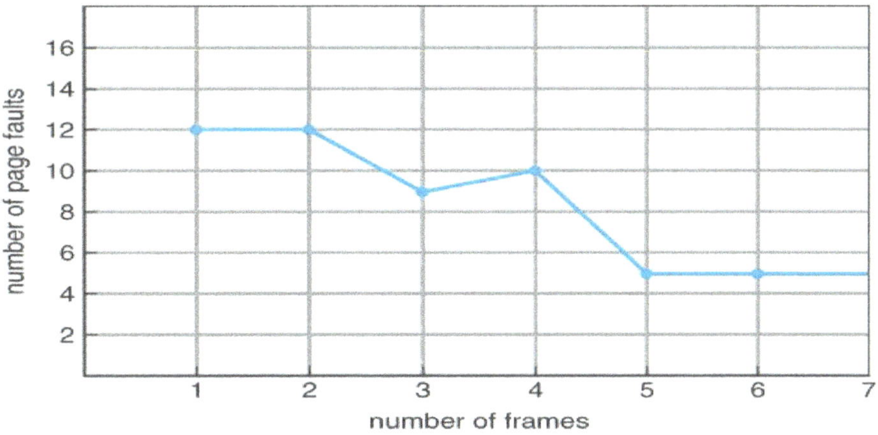

3.7.2.2 Optimal Page Replacement

- The discovery of Belady's anomaly lead to the search for an **optimal page-replacement algorithm**, which is simply that which yields the lowest of all possible page-faults, and which does not suffer from Belady's anomaly.
- Such an algorithm does exist, and is called **OPT or MIN.** This algorithm is simply "Replace the page that will not be used for the longest time in the future."
- **For example**, by applying OPT to the same reference string used for the FIFO example, the minimum number of possible page faults is 9. Since 6 of the page-faults are unavoidable (the first reference to each new page), FIFO can be shown to require 3 times as many (extra) page faults as the optimal algorithm.

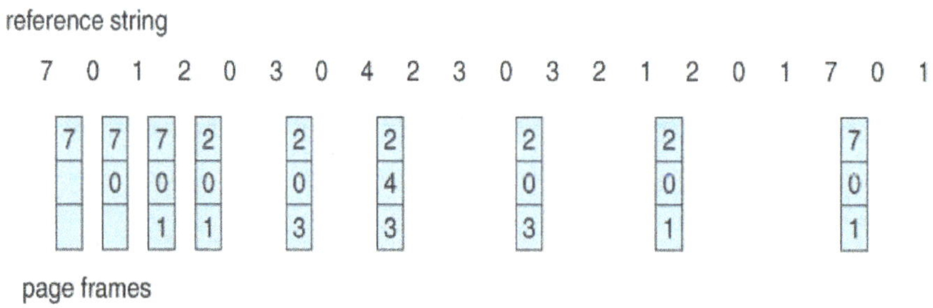

Figure 3.34 Optimal page-replacement algorithm

3.7.2.3 LRU Page Replacement

- The prediction behind **LRU,** the **Least Recently Used,** and algorithm is that the page that has not been used in the longest time is the one that will not be used again in the near future.
- Some view LRU as analogous to OPT, except looking backwards in time instead of forwards. (OPT has the interesting property that for any reference string S and its reverse R, OPT will generate the same number of page faults for S and for R. It turns out that LRU has this same property.)
- LRU yields 12 page faults, (as compared to 15 for FIFO and 9 for OPT.)

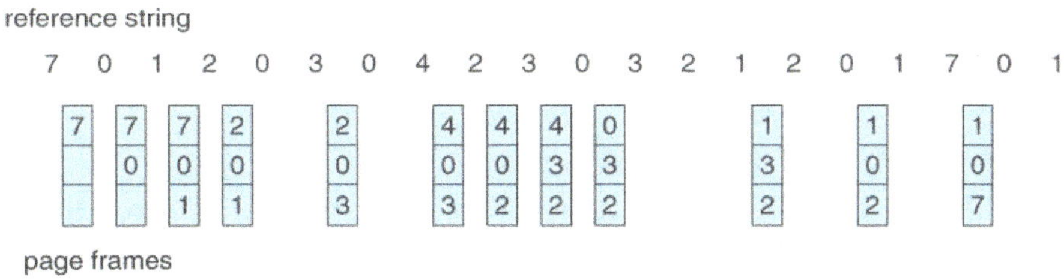

Figure 3.34 LRU page-replacement algorithm.

LRU is considered a good replacement policy, and is often used. The problem is how exactly to implement it. There are **two simple approaches** commonly used:

1. **Counters.** Every memory access increments a counter and the current value of this counter is stored in the page table entry for that page. Then finding the LRU page involves simple searching the table for the page with the smallest counter value.
2. **Stack.** Whenever a page is accessed, pull that page from the middle of the stack and place it on the top. The LRU page will always be at the bottom of the stack.

Both implementations of LRU require hardware support, either for incrementing the counter or for managing the stack, as these operations must be performed for **every** memory access.

- Neither LRU or OPT exhibit Belady's anomaly. Both belong to a class of page-replacement algorithms called **stack algorithms,** which can never exhibit Belady's anomaly.
- A stack algorithm is one in which the pages kept in memory for a frame set of size N will always be a subset of the pages kept for a frame size of N + 1.

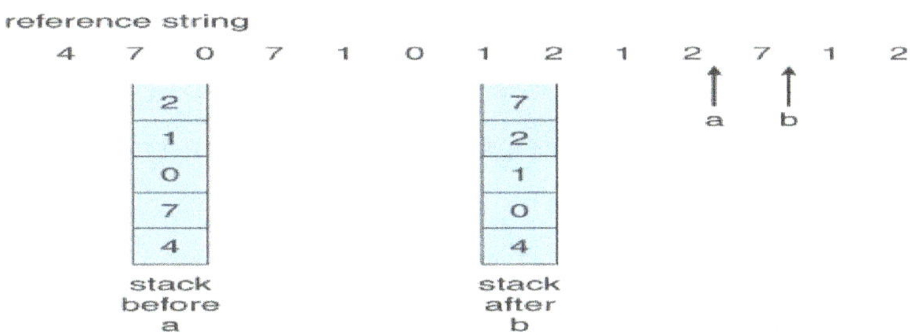

Figure 3.35 Use of a stack to record the most recent page references.

3.7.2.4 LRU-Approximation Page Replacement

- Many systems provide a **reference bit** for every entry in a page table, which is set anytime that page is accessed.
- Initially all bits are set to zero, and they can also all be cleared at any time.
- One bit of precision is enough to distinguish pages that have been accessed since the last clear from those that have not, but does not provide any finer grain of detail.

3.7.2.4. Additional-Reference-Bits Algorithm

Finer grain is possible by storing the most recent 8 reference bits for each page in an 8-bit byte in the page table entry, which is interpreted as an unsigned int.

- At periodic intervals (clock interrupts), the OS takes over, and right-shifts each of the reference bytes by one bit.
- The high-order (leftmost) bit is then filled in with the current value of the reference bit, and the reference bits are cleared.
- At any given time, the page with the smallest value for the reference byte is the LRU page.
- Obviously the specific number of bits used and the frequency with which the reference byte is updated are adjustable, and are tuned to give the fastest performance on a given hardware platform.

3.7.2.4.2 Second-Chance Algorithm

The **second chance algorithm** is essentially a FIFO, except the reference bit is used to give pages a second chance at staying in the page table.

- When a page must be replaced, the page table is scanned in a FIFO (circular queue) manner.
- If a page is found with its reference bit not set, then that page is selected as the next victim.
- If, however, the next page in the FIFO does have its reference bit set, then it is given a second chance:
  - ➢ The reference bit is cleared, and the FIFO search continues.

- ➢ If some other page is found that did not have its reference bit set, then that page will be selected as the victim, and this page (the one being given the second chance) will be allowed to stay in the page table.
- ➢ If there are no other pages that do not have their reference bit set, then this page will be selected as the victim when the FIFO search circles back around to this page on the second pass.

- If all reference bits in the table are set, then second chance degrades to FIFO, but also requires a complete search of the table for every page-replacement.
- As long as there are some pages whose reference bits are not set, then any page referenced frequently enough gets to stay in the page table indefinitely.
- This algorithm is also known as the **clock** algorithm, from the hands of the clock moving around the circular queue.

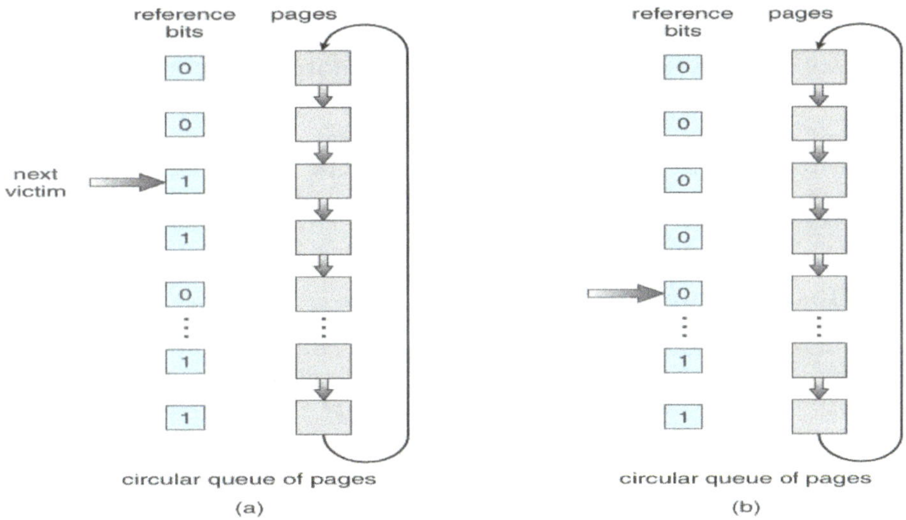

Figure 3.36 Second-chance (clock) page-replacement algorithm.

3.7.2.4.3 Enhanced Second-Chance Algorithm

The **enhanced second chance algorithm** looks at the reference bit and the modify bit (dirty bit) as an ordered page, and classifies pages into one of four classes:

- (0, 0) - Neither recently used nor modified.
- (0, 1) - Not recently used, but modified.
- (1, 0) - Recently used, but clean.
- (1, 1) - Recently used and modified.

This algorithm searches the page table in a circular fashion (in as many as four passes), looking for the first page it can find in the lowest numbered category. i.e. it first makes a pass looking for a (0, 0), and then if it can't find one, it makes another pass looking for a (0, 1), etc.

- Difference: It is the preference for replacing clean pages if possible.

3.7.2.5 Counting-Based Page Replacement

There are several algorithms based on counting the number of references that have been made to a given page, such as:

3.7.2.5.1 Least Frequently Used (LFU)

Replace the page with the lowest reference count. A problem can occur if a page is used frequently initially and then not used any more, as the reference count remains high.

A solution to this problem is to right-shift the counters periodically, yielding a time-decaying average reference count.

3.7.2.5.2 Most Frequently Used (MFU)

Replace the page with the highest reference count.

Logic: The pages that have already been referenced a lot have been in the system a long time, and probably done with them, whereas pages referenced only a few times have only recently been loaded, and still need them.

In general counting-based algorithms are not commonly used, as their implementation is expensive and they do not approximate OPT well.

3.7.2.6 Page-Buffering Algorithms

There are a number of page-buffering algorithms to improve overall performance.

- Maintain a certain **minimum number of free frames** at all times. When a page-fault occurs, go ahead and allocate one of the free frames from the free list first, to get the requesting process up and running again as quickly as possible, and then select a victim page to write to disk and free up a frame as a second step.

- **Keep a list of modified pages**, and when the I/O system is otherwise idle, have it write these pages out to disk, and then clear the modify bits, thereby increasing the chance of finding a "clean" page for the next potential victim.
- **Keep a pool of free frames**, but remember what page was in it before it was made free. Since the data in the page is not actually cleared out when the page is freed, it can be made an active page again without having to load in any new data from disk.

3.7.3 Applications and Page Replacement

- All of these algorithms have OS guessing about future page access
- Some applications have better knowledge – i.e. databases
- Memory intensive applications can cause double buffering
  - ➢ OS keeps copy of page in memory as I/O buffer
  - ➢ Application keeps page in memory for its own work
- Operating system can given direct access to the disk, getting out of the way of the applications
  - ➢ **Raw disk** mode
- Bypasses buffering, locking, etc

3.8 ALLOCATION OF FRAMES

Minimum Number of Frames

Reasons:

If an instruction (and its operands) spans a page boundary, then **multiple pages could be needed** just for the instruction fetch.

Memory references in an instruction touch more pages, and if those memory locations can span page boundaries, then multiple pages could be needed for operand access also.

The **worst case** involves indirect addressing, particularly where multiple levels of indirect addressing are allowed. Left unchecked, a pointer to a pointer to a pointer to a pointer to a . . . could theoretically touch every page in the virtual address space in a single machine instruction, requiring every virtual page be loaded into physical memory simultaneously.

So place a limit (say 16) on the number of levels with a counter initialized to the limit and decremented with every level of indirection in an instruction. If the counter reaches zero, then an "excessive indirection" trap occurs. This requires a minimum frame allocation of 17 per process.

3.8.1 Allocation Algorithms

i) Equal Allocation - If there are m frames available and n processes to share them, each process gets m / n frames, and the leftovers are kept in a free-frame buffer pool.

ii) Proportional Allocation - Allocate the frames proportionally to the size of the process, relative to the total size of all processes. So if the size of process i is S_i, and S is the sum of all S_i, then the allocation for process P_i is a_i = m * S_i / S.

Variations on proportional allocation could consider priority of process rather than just their size.

All allocations fluctuate over time as the number of available free frames, m, fluctuates, and all are also subject to the constraints of minimum allocation.

iii) Global versus Local Allocation - With **local replacement**, the number of pages allocated to a process is fixed, and page replacement occurs only amongst the pages allocated to this process.

It allows processes to better control their own page fault rates, and leads to more consistent performance of a given process over different system load levels.

With **global replacement**, any page may be a potential victim, whether it currently belongs to the process seeking a free frame or not.

It is more efficient, and is the more commonly used approach.

iv) Non-Uniform Memory Access

It is assumed that all memory is equivalent, or at least has equivalent access times.

CPUs can access memory that is physically located on the same board much faster than the memory on the other boards.

The basic **solution** is akin to processor affinity - trying to schedule processes on the same CPU will minimize cache misses, also trying to allocate memory for those processes on the same boards will minimize access times.

The presence of threads complicates the picture, especially when the threads get loaded onto different processors.

Solaris uses an **lgroup** as a solution, in a hierarchical fashion based on relative latency.

For example, all processors and RAM on a single board would probably be in the same lgroup.

Memory assignments are made within the same lgroup if possible or to the next nearest lgroup otherwise. (Where "nearest" is defined as having the lowest access time).

3.9 Thrashing

A process that is spending more time paging than executing is said to be **thrashing.**

Cause of Thrashing

- When memory is filled up and processes started spending lots of time waiting for their pages to page in, then CPU utilization would lower, causing the schedule to add in even more processes and the system would grind to a halt.
- **Local page replacement policies can prevent** one thrashing process from taking pages away from other processes.

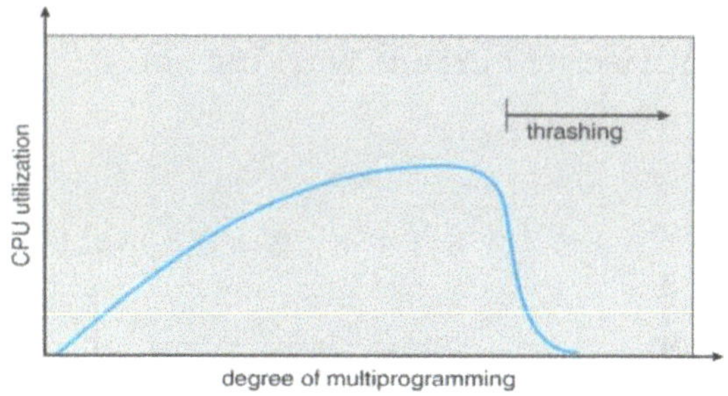

Figure 3.37 Thrashing

- To prevent thrashing, must provide processes with as many frames as possible.
- The **locality model** notes that processes typically access memory references in a given **locality,** making lots of references to the same general area of memory before moving periodically to a new locality.
- If many frames are involved in the current locality, then page faulting would occur primarily on switches from one locality to another.

Figure 3.38 Locality in a memory-reference pattern.

3.9.1 Working-Set Model

- The **working set model** is based on the concept of locality, and defines a **working set window**, of length **delta.** Whatever pages are included in the most recent delta page references are said to be in the processes working set window, and comprise its current working set.

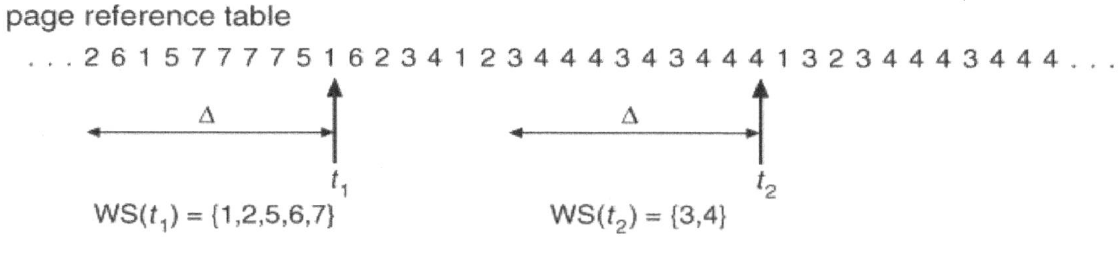

Figure 3.39 Working-set model.

- The selection of delta is critical to the success of the working set model
  - If it is too small then it does not encompass all of the pages of the current locality.
  - if it is too large, then it encompasses pages that are no longer being frequently accessed.
- The total demand, D = sum of the sizes of the working sets for all processes.

- If D > total number of available frames, then at least one process is thrashing, because there are not enough frames available to satisfy its minimum working set.
- If D < the currently available frames, then additional processes can be launched.

- The hard part of the working-set model is keeping track of what pages are in the current working set, since every reference adds one to the set and removes one older page.
- An approximation can be made using reference bits and a timer that goes off after a set interval of memory references:
  - For example, suppose that the timer is set to go off after every 5000 references (by any process), and can store two additional historical reference bits in addition to the current reference bit.
  - Every time the timer goes off, the current reference bit is copied to one of the two historical bits, and then cleared.
  - If any of the three bits is set, then that page was referenced within the last 15,000 references, and is considered to be in that processes reference set.
  - Finer resolution can be achieved with more historical bits and a more frequent timer, at the expense of greater overhead.

3.9.2 Page-Fault Frequency

- A more direct approach is to recognize that what really want to control is the page-fault rate, and to allocate frames based on this directly measurable value.
- If the page-fault rate exceeds a certain upper bound then that process needs more frames, and if it is below a given lower bound, then it can afford to give up some of its frames to other processes.

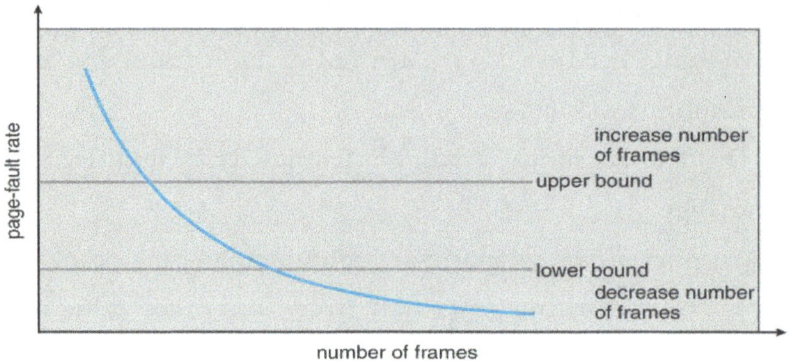

Figure 3.40 Page-fault frequency.

- There is a direct relationship between the page-fault rate and the working-set, as a process moves from one locality to another:

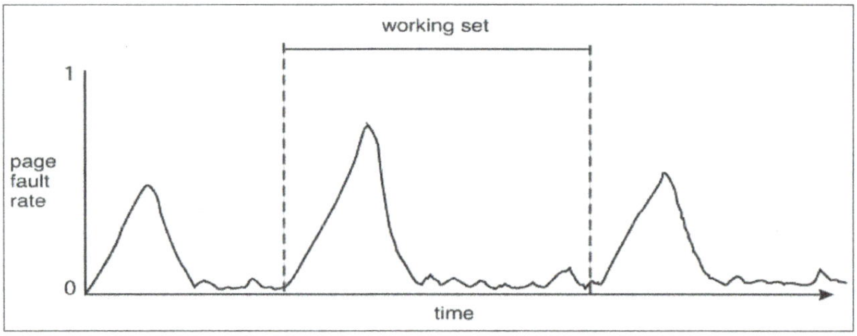

Figure 3.40 Unnumbered side bar in Ninth Edition

3.10 Allocating Kernel Memory

There are several classic algorithms in place for allocating kernel memory structures.

Buddy System

- The Buddy **System** allocates memory using a **power of two allocator**.
- Under this scheme, memory is always allocated as a power of 2 (4K, 8K, 16K, etc), rounding up to the next nearest power of two if necessary.
- If a block of the correct size is not currently available, then one is formed by splitting the next larger block in two, forming two matched buddies. (And if that larger size is not available, then the next largest available size is split, and so on).

- Feature- if the address of a block is exclusively ORed with the size of the block, the resulting address is the address of the buddy of the same size, which allows for fast and easy **coalescing** of free blocks back into larger blocks.
    - ➢ Free lists are maintained for every size block.
    - ➢ If the necessary block size is not available upon request, a free block from the next largest size is split into two buddies of the desired size. (Recursively splitting larger size blocks if necessary.)
    - ➢ When a block is freed, its buddy's address is calculated, and the free list for that size block is checked to see if the buddy is also free. If it is, then the two buddies are coalesced into one larger free block, and the process is repeated with successively larger free lists.

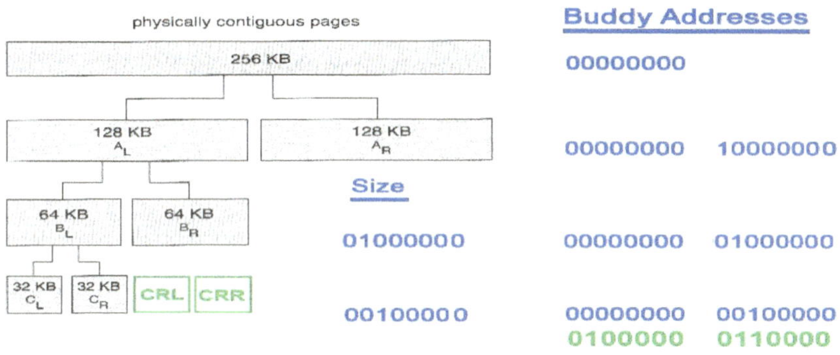

Figure 3.42 Slab Allocation

- **Slab Allocation** allocates memory to the kernel in chunks called **slabs**, consisting of one or more contiguous pages. The kernel then creates separate caches for each type of data structure it might need from one or more slabs. Initially the caches are marked empty, and are marked full as they are used.
- New requests for space in the cache is first granted from empty or partially empty slabs, and if all slabs are full, then additional slabs are allocated.
- For example if a particular structure were 512 bytes long, space for them would be allocated in groups of 8 using 4K pages. If the structure were 3K, then space for 4 of them could be allocated at one time in a slab of 12K using three 4K pages.
- Benefits
    - lack of internal fragmentation and
    - fast allocation of space for individual structures

New in 9th Edition: Linux SLOB and SLUB allocators replace SLAB

- SLOB, Simple List of Blocks, maintains 3 linked lists of free blocks - small, medium, and large - designed for (imbedded) systems with limited amounts of memory.
- SLUB modifies some implementation issues for better performance on systems with large numbers of processors.

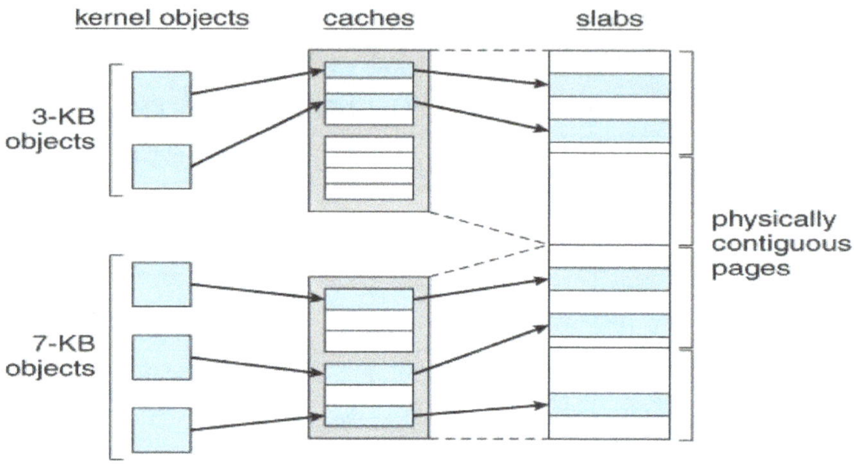

Figure 3.43 Slab allocations

3.11 OPERATING-SYSTEM EXAMPLES

3.11.1 Windows

- Windows uses demand paging with **clustering,** meaning they page in multiple pages whenever a page fault occurs.
- The working set minimum and maximum are normally set at 50 and 345 pages respectively.
- Free pages are maintained on a free list, with a minimum threshold indicating when there are enough free frames available.
- If a page fault occurs and the process is below their maximum, then additional pages are allocated. Otherwise some pages from this process must be replaced, using a local page replacement algorithm.
- If the amount of free frames falls below the allowable threshold, then **working set trimming** occurs, taking frames away from any processes which are above

their minimum, until all are at their minimums. Then additional frames can be allocated to processes that need them.

- The algorithm for selecting victim frames depends on the type of processor:
    - On single processor 80x86 systems, a variation of the clock (second chance) algorithm is used.
    - On Alpha and multiprocessor systems, clearing the reference bits may require invalidating entries in the TLB on other processors, which is an expensive operation. In this case Windows uses a variation of FIFO.

3.11.2 Solaris

- Solaris maintains a list of free pages, and allocates one to a faulting thread whenever a fault occurs. It is therefore imperative that a minimum amount of free memory be kept on hand at all times.
- Solaris has a parameter, lotsfree, usually set at 1/64 of total physical memory. Solaris checks 4 times per second to see if the free memory falls below this threshhold, and if it does, then the **pageout** process is started.
- Pageout uses a variation of the clock (second chance) algorithm, with two hands rotating around through the frame table.
    - The first hand clears the reference bits, and
    - The Second hand comes by afterwards and checks them. Any frame whose reference bit has not been reset before the second hand gets there gets paged out.
- The Pageout method is adjustable by the distance between the two hands, (the **handspan**), and the speed at which the hands move.
- For example, if the hands each check 100 frames per second, and the handspan is 1000 frames, then there would be a 10 second interval between the time when the leading hand clears the reference bits and the time when the trailing hand checks them.
- The speed of the hands is usually adjusted according to the amount of free memory.
- **Slowscan** is usually set at 100 pages per second, and **fastscan** is usually set at the smaller of 1/2 of the total physical pages per second and 8192 pages per second.

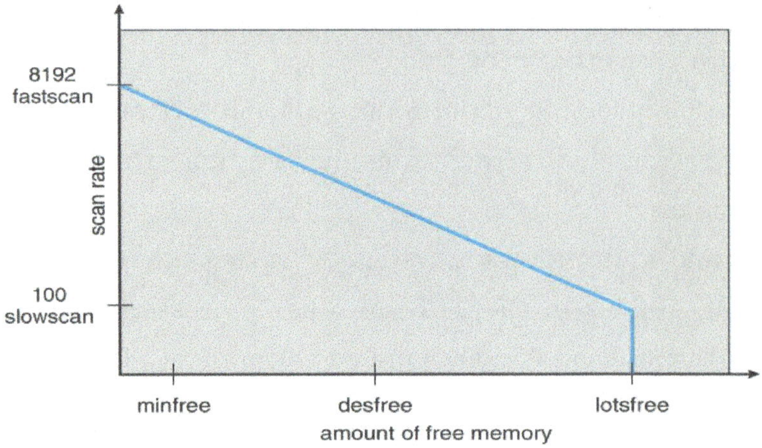

Figure 3.44 Solaris page scanner.

- Solaris also maintains a cache of pages that have been reclaimed but which have not yet been overwritten, as opposed to the free list which only holds pages whose current contents are invalid.
- If one of the pages from the cache is needed before it gets moved to the free list, then it can be quickly recovered.
- Normally pageout runs 4 times per second to check if memory has fallen below lotsfree. However if it falls below desfree, then pageout will run at 100 times per second in an attempt to keep at least desfree pages free. If it is unable to do this for a 30-second average, then Solaris begins swapping processes, starting preferably with processes that have been idle for a long time.
- If free memory falls below minfree, then pageout runs with every page fault.

3

STORAGE MANAGEMENT

REVIEW QUESTIONS

PART-A

1. **What is meant by base and limit registers?**
   - Each process has a **separate memory space** to **protects the processes** from each other,
   - To separate memory spaces, the range of legal addresses that the process may access need to be determined and to ensure that the process can access only these legal addresses.
   - This protection is provided **by using two registers**, usually **a base and a limit** i.e., a pair of base and limit registers define the logical address space

2. **What are the three different stages of address binding?**
   - **Compile time**: If memory location known a priori, absolute code can be generated; must recompile code if starting location changes
   - **Load time**: Must generate relocatable code if memory location is not known at compile time
   - **Execution time**: Binding delayed until run time if the process can be moved during its execution from one memory segment to another

3. **Define logical and physical address of a system.**
    - **Logical address** – generated by the CPU; also referred to as virtual address
    - **Physical address** – address seen by the memory unit and is loaded into Memory Address Register (MAR)

4. **What is the purpose of a MMU**
    - Hardware device that at run time maps **virtual to physical address**

5. **With neat diagram brief about Dynamic loading and relocation register.**
    - CPU produces logical address
    - Logical address is added with the relocation register value to get the physical address

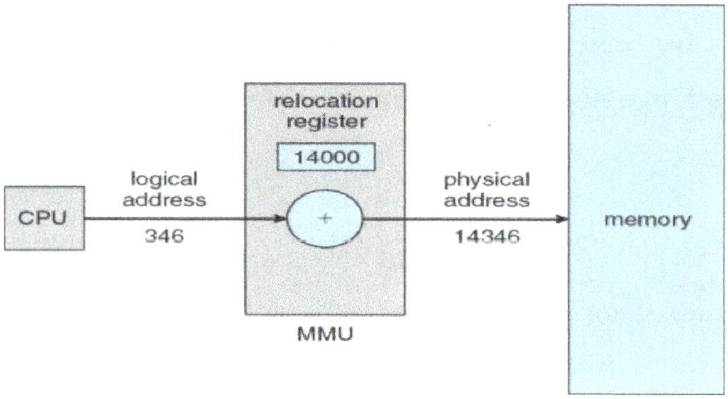

6. **What is static and dynamic linking?**
    - **Static linking** – system libraries and program code combined by the loader into the binary program image
    - **Dynamic linking** –linking postponed until execution time

7. **What are the three dynamic allocation strategies?**
    - **First-fit**: Allocate the **first** hole that is big enough
    - **Best-fit**: Allocate the **smallest** hole that is big enough; must search entire list, unless ordered by size
        - Produces the smallest leftover hole
    - **Worst-fit**: Allocate the **largest** hole; must also search entire list
        - Produces the largest leftover hole

- First-fit and best-fit better than worst-fit in terms of speed and storage utilization

8. What are the types of Fragmentation?

- **Internal fragmentation**
- **External fragmentation**

9. What is meant by internal fragmentation?

Allocated memory may be slightly larger than requested memory; this size difference is memory internal to a partition, but not being used

10. What is meant by external fragmentation?

Total memory space exists to satisfy a request, but it is not

11. Give the solution to External fragmentation.

- Reduce external fragmentation by **compaction**
    - ➢ Shuffle memory contents to place all free memory together in one large block
    - ➢ Compaction is possible only if relocation is dynamic, and is done at execution time

12. What is meant by Segmentation and segment table?

- A segment is a logical unit such as: main program, procedure, function, method, object, local variables, global variables, common block, stack , symbol table, arrays
- Segment table – maps two-dimensional physical addresses; each table entry has:
    - ➢ base – contains the starting physical address where the segments reside in memory
    - ➢ limit – specifies the length of the segment
    - ➢ Segment-table base register (STBR) points to the segment table's location in memory

13. What is STLR and how segment table is protected?

Segment table length register indicates number of segments used by a program; segment number s is legal if s < STLR

- Protection
  - ➢ With each entry in segment table associate:
    - ❖ validation bit = 0 ⇒ illegal segment
    - ❖ read/write/execute privileges
- Protection bits associated with segments; code sharing occurs at segment level

14. List out the advantages of using paging.

- Physical address space of a process can be non-contiguous; process is allocated physical memory whenever the latter is available
  - ➢ Avoids external fragmentation
  - ➢ Avoids problem of varying sized memory chunks

15. In a system, page size is 2048 bytes and the process needs 72, 766 bytes of size. Find out whether internal fragmentation occurs or not?

Page size = 2,048 bytes

- Process size = 72,766 bytes
- 35 pages + 1,086 bytes
- Internal fragmentation of 2,048 - 1,086 = 962 bytes

16. What is meant by TLB?

The two memory access problem can be solved by the use of a special fast-lookup hardware cache called **associative memory** or **translation look-aside buffers (TLBs)**

Some TLBs store **address-space identifiers (ASIDs)** in each TLB entry – uniquely identifies each process to provide address-space protection for that process

17. Define Effective Access time.

Depends on TLB hit and miss ratio.

If TLB hit ratio is 80% and Memory needs 100 ns then

Effective Access Time (EAT) = 0.80 x 100 + 0.20 x 200 = 120ns

18. How are memory protection implemented?

- **Valid-invalid** bit attached to each entry in the page table:
  - "valid" indicates that the associated page is in the process' logical address space, and is thus a legal page
  - "invalid" indicates that the page is not in the process' logical address space

19. Define LDT and GDT.

The Intel IA-32 Architecture supports for both segmentation and paging

Divided into two partitions

- First partition of up to 8 K segments are private to process (kept in **local descriptor table (LDT**))
- Second partition of up to 8K segments shared among all processes (kept in **global descriptor table (GDT**))

20. Define virtual memory.

Separation of user logical memory from physical memory

- Only part of the program needs to be in memory for execution
- Logical address space can therefore be much larger than physical address space
- Allows address spaces to be shared by several processes
- Allows for more efficient process creation
- More programs running concurrently
- Less I/O needed to load or swap processes

21. How virtual memory be implemented?

- Virtual memory can be implemented via:
  - Demand paging
  - Demand segmentation

22. What is meant by Demand paging?

- Brings the entire process into memory at load time
- i.e., bring a page into memory only when it is needed
  - Less I/O needed, no unnecessary I/O

- ➢ Less memory needed
- ➢ Faster response
- ➢ More users

23. Define Lazy swapper.

- **Lazy swapper** – never swaps a page into memory unless page will be needed
  - ➢ Swapper that deals with pages is a **pager**

24. What is a page fault?

If there is a reference to a page, first reference to that page will trap to operating system: **page fault**

25. List out the steps of page fault recovery.

1. Operating system looks at another table to decide:
   - Invalid reference ⇒ abort
   - Just not in memory
2. Find free frame
3. Swap page into frame via scheduled disk operation
4. Reset tables to indicate page now in memory Set validation bit = **v**
5. Restart the instruction that caused the page fault

26. List out the steps of page replacement.

1. Find the location of the desired page on disk
2. Find a free frame:
   - ➢ If there is a free frame, use it
   - ➢ If there is no free frame, use a page replacement algorithm to select a **victim frame**
   - ➢ Write victim frame to disk if dirty
3. Bring the desired page into the (newly) free frame; update the page and frame tables
4. Continue the process by restarting the instruction that caused the trap

27. List out Page Replacement Algorithms.

1. First-In-First-Out (FIFO) Algorithm
2. Optimal Algorithm
3. Least Recently Used (LRU) Algorithm
4. LRU Approximation Algorithms
5. Counting Algorithms
6. Page-Buffering Algorithms

28. Give the different methods to allocate frames.

1. Fixed allocation
2. Priority Allocation
3. Global Vs Local Allocation
4. Non Uniform Memory Allocation

29. What is meant by thrashing?

Thrashing ≡ a process is busy swapping pages in and out

30. What are the methods used to allocate kernel memory?

1. Buddy System Allocator
2. Slab Allocator

31. List out he Examples for OS.

1. Windows
2. Solaris

PART-B

1. Explain in detail about Segmentation (8)
2. Explain about paging in detail. (16)
3. Explain about paging with Intel 32, 64 bit architecture. (8)
4. Discuss in detail about Demand Paging. (10)
5. Explain in detail about Page Replacement Algorithms with Examples. (16)
6. Write short notes the following
   - ➢ Different methods of frame allocation (8)
   - ➢ Thrashing (8)
   - ➢ Allocation of Kernel memory (8)

4

I/O SYSTEMS

4.1 OVERVIEW OF MASS-STORAGE STRUCTURE

4.1.1 Magnetic Disks

Magnetic disks provide the bulk of secondary storage for modern computer systems. Conceptually, disks are relatively simple.

Each disk platter has a flat circular shape, like a CD and the two surfaces of a platter are covered with a magnetic material where the information is stored by recording it magnetically on the platters.

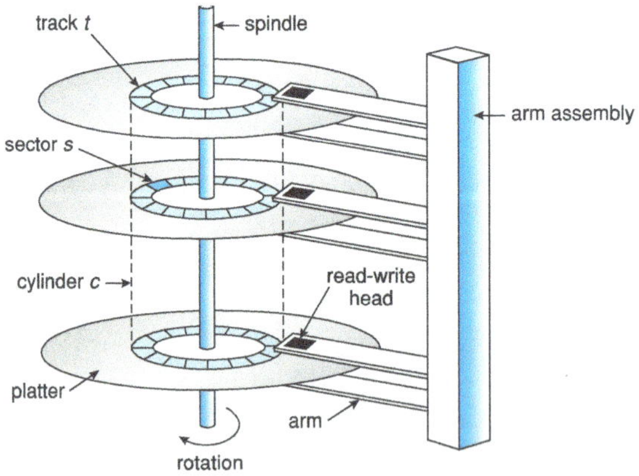

Figure 4.1 Moving-head disk mechanism

Description:

The disc platter is a flat circular plate like CD & the two surface of a platter are covered with magnetic materials in which the information is stored by recording magnetically.

The surface of a platter is logically divided into circular **tracks**, which are subdivided into **sectors**. The heads are attached to disk arm that moves all the heads as a unit.

There may be thousands of concentric cylinders in a disk drive, and each track may contain hundreds of sectors.

Most drives rotate 60 to 250 times per second, specified in terms of rotations per minute (RPM). Common drives spin at 5,400, 7,200, 10,000, and 15,000RPM.

Disk speed has two parts.

- The **transfer rate** is the rate at which data flow between the drive and the computer.
- The **positioning time**, **or random-access time,** consists of two parts:
  - ➢ The time necessary to move the disk arm to the desired cylinder- **seek time**
  - ➢ The time necessary for the desired sector to rotate to the disk head- **rotational latency.**

Because the disk head flies on an extremely thin cushion of air (measured in microns), there is a danger that the head will make contact with the disk surface.

Although the disk platters are coated with a thin protective layer, the head will sometimes damage the magnetic surface. This accident is called a **head crash**. (cannot be repaired; the entire disk must be replaced).

A disk drive is attached to a computer by a set of wires called an **I/O bus**.

There are **several kinds of buses** available that includes advanced technology attachment (ATA), serialATA (SATA), eSATA, universal serial bus (USB), and fibre channel (FC).

The data transfers on a bus are carried out by special electronic processors called **controllers**.

- The **host controller** is the controller at the computer end of the bus.
- A **disk controller** is built into each disk drive.

To **perform a disk I/O operation**, the computer places a command into the host controller, and then sends the command via messages to the disk controller, and the disk controller operates the disk-drive hardware to carry out the command.

Disk controllers usually have a built-in cache. Data transfer at the disk drive happens between the cache and the disk surface, and data transfer to the host, occurs between the cache and the host controller.

4.1.2 Solid state Disks

- Nonvolatile memory used like a hard drive
  - Many technology variations from DRAM with a battery to allow it to maintain its state in a power failure through flash-memory technologies like single level cell (SLC) and multi level cell (MLC)
- more reliable than HDDs
- More expensive per MB
- Maybe have shorter life span
- Less capacity
- But much faster
- Busses can be too slow ->
- Hence SSDs connected directly to system bus, PCI for example
- No moving parts, so no seek time or rotational latency

4.1.3 Magnetic tapes

- Was early secondary-storage medium
  - Evolved from open spools to cartridges
- Relatively permanent and holds large quantities of data
- Access time slow
- Random access ~1000 times slower than disk
- Mainly used for backup, storage of infrequently-used data, transfer medium between systems
- Kept in spool and wound or rewound past read-write head
- Once data under head, transfer rates comparable to disk
  - 140MB/sec and greater
- 200GB to 1.5TB typical storage
- Common technologies are LTO-{3,4,5} and T10000

4.2 DISK SCHEDULING

One of the responsibilities of the OS is to use the hardware efficiently. The disk drives have fast access time and large disk bandwidth.

For magnetic disks, the **access time** has **two major components**:

- The **seek time** is the time for the disk arm to move the heads to the cylinder containing the desired sector.
- The **rotational latency** is the additional time for the disk to rotate the desired sector to the disk head.

The **disk bandwidth** is the total number of bytes transferred, divided by the total time between the first request for service and the completion of the last transfer. To improve both the access time and the bandwidth the disk I/O requests are serviced. Whenever a process needs I/O to or from the disk, it issues a system call to the OS.

The request specifies several **pieces of information**:

- Whether this operation is input or output
- What the disk address for the transfer is
- What the memory address for the transfer is
- What the number of sectors to be transferred is

If the desired disk drive and controller are available, the request can be serviced immediately else it is placed in the queue of pending requests for that drive. There are several **disk-scheduling algorithms** to be used.

4.2.1 FCFS Scheduling

The simplest form of disk scheduling is the first-come, first-served (FCFS) algorithm. This algorithm does not provide the fastest service.

Consider, for example, a disk queue with requests for I/O to blocks on cylinders. The scheduling algorithm with a request queue (0-199)→ 98, 183, 37, 122, 14, 124, 65, 67

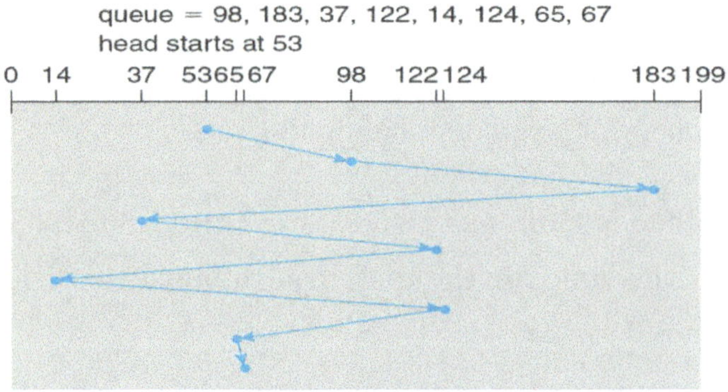

Figure 4.2 FCFS disk scheduling.

If the disk head is initially at cylinder 53, it will first move from 53 to 98, then to 183, 37, 122, 14, 124, 65, and finally to 67, for a **total head movement of 640 cylinders**.

The wild swing from 122 to 14 and then back to 124 **illustrates the problem** with this schedule.

If the requests for cylinders 37 and 14 could be serviced together, before or after the requests for 122 and 124, the total head movement could be decreased substantially, and performance could be thereby improved.

4.2.2 SSTF Scheduling

The SSTF algorithm selects the request with the least seek time from the current head position. In other words, SSTF chooses the pending request closest to the current head position.

Example: In the request queue, the closest request to the initial head position (53) is at cylinder 65. Once it is at cylinder 65, the next closest request is at cylinder 67. From there, the request at cylinder 37 is closer than the one at 98, so 37 is served next. Continuing the service of request at cylinder 14, then 98, 122, 124, and finally 183.

This scheduling method results in a **total head movement of only 236 cylinders**—little more than **one-third of the distance needed for FCFS** scheduling of this request queue.

SSTF scheduling may **cause starvation** of some requests.

Example: The requests may arrive at any time. Suppose that there are two requests in the queue, for cylinders 14 and 186, and while the request from 14 is being serviced, a new request near 14 arrives. This new request will be serviced next, making the request at 186 wait. While this request is being serviced, another request close to 14 could arrive. In theory, a continual stream of requests near one another could cause the request for cylinder 186 to **wait indefinitely**.

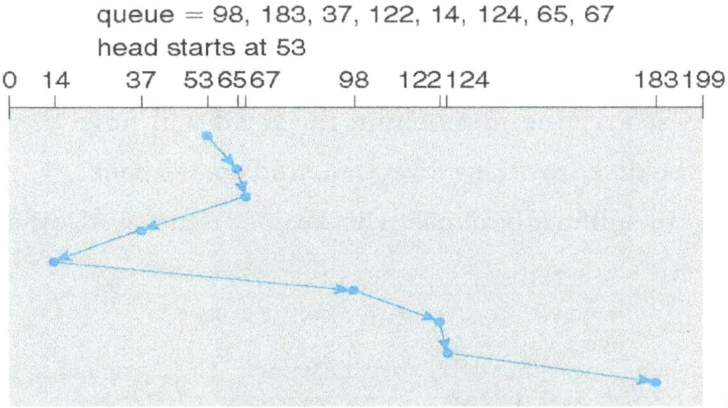

3

Figure 4.3 SSTF disk scheduling.

This becomes increasingly as the pending-request queue grows longer.

The SSTF algorithm is **not optimal**. In the example, the **better option** is by moving the head from 53 to 37, even though the latter is not closest, and then to 14, before turning around to service 65, 67, 98, 122, 124, and 183. This strategy reduces the **total head movement to 208 cylinders**.

4.2.3 SCAN Scheduling

In the SCAN algorithm, the disk arm starts at one end of the disk and moves toward the other end, servicing requests as it reaches each cylinder, until it gets to the other end of the disk. At the other end, the direction of head movement is reversed, and servicing continues.

The head continuously scans back and forth across the disk. The SCAN algorithm is sometimes called the **elevator algorithm**, since the disk arm behaves just like an elevator in a building, first servicing all the requests going up and then reversing to service requests the other way.

Example: Before applying SCAN to schedule the requests on cylinders 98, 183, 37, 122, 14, 124, 65, and 67, it is needed to know the direction of head movement in addition to the head's current position.

Assuming that the disk arm is moving toward 0 and that the initial head position is again 53, the head will next service 37 and then 14. At cylinder 0, the arm will reverse and will move toward the other end of the disk, servicing the requests at 65, 67, 98, 122, 124, and 183.

If a request arrives in the queue just in front of the head, it will be serviced almost immediately;

And if a request arriving just behind the head will have to wait until the arm moves to the end of the disk, reverses direction, and comes back.

But if the requests are uniformly dense, the largest density at other end of disk and those wait the longest.

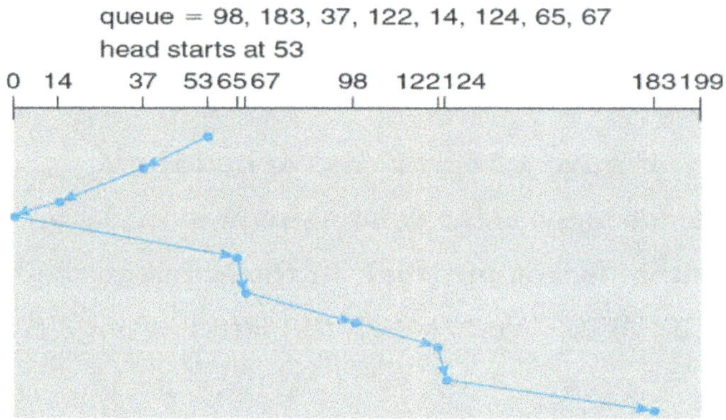

Figure 4.4 SCAN disk scheduling.

4.2.4 C-SCAN Scheduling

Circular SCAN(C-SCAN) scheduling is a variant of SCAN designed to provide a **more uniform wait time.**

Like SCAN, C-SCAN moves the head from one end of the disk to the other, servicing requests along the way. When the head reaches the other end, it immediately returns to the beginning of the disk without servicing any requests on the return trip. The C-SCAN scheduling algorithm essentially treats the cylinders as a circular list that wraps around from the final cylinder to the first one.

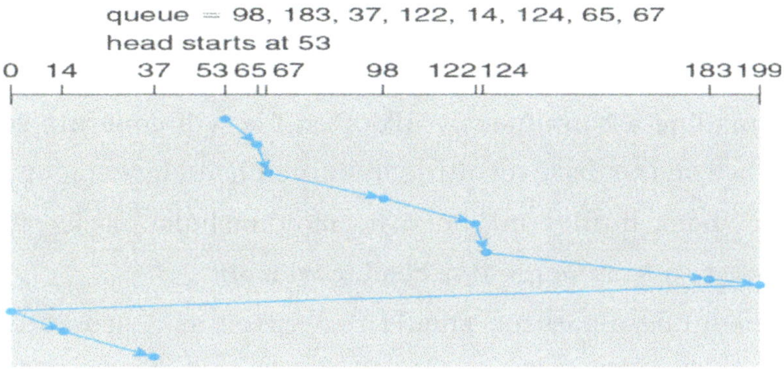

Figure 4.5 C-SCAN disk scheduling.

4.2.5 LOOK Scheduling

Both SCAN and C-SCAN move the disk arm across the full width of the disk. The arm goes as the final request in each direction. Then, it reverses direction immediately, without going all the way to the end of the disk.

Versions of SCAN and C-SCAN that follow this pattern are called LOOK and C-LOOK scheduling, because they look for a request before continuing to move in a given direction.

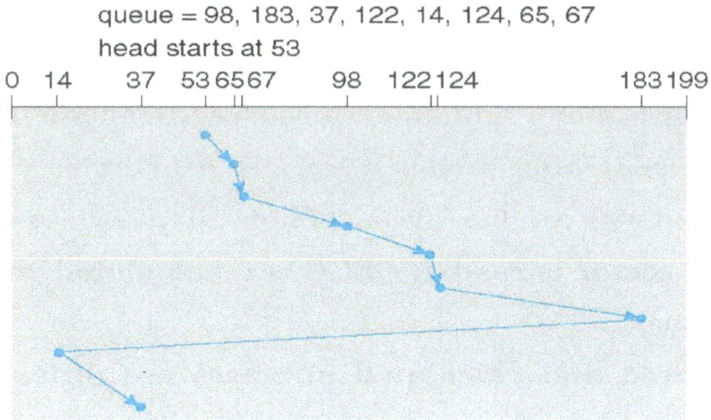

Figure 4.6 C-LOOK disk scheduling.

Selection of a Disk-Scheduling Algorithm

SSTF is common and has a natural appeal because it increases performance over FCFS.

SCAN and C-SCAN perform better for systems that place a heavy load on the disk, because they are less likely to cause a starvation problem.

The performance depends heavily on the number and types of requests.

Requests for disk service can be greatly influenced by the file-allocation method.

A program reading a contiguously allocated file will generate several requests that are close together on the disk, resulting in limited head movement.

A linked or indexed file, in contrast, may include blocks that are widely scattered on the disk, resulting in greater head movement.

The disk-scheduling algorithm should be written as a separate module of the operating system, allowing it to be replaced with a different algorithm if necessary

4.3 DISK MANAGEMENT

- A new magnetic disk is a blank slate with a platter of a magnetic recording material. Dividing a disk into sectors that the disk controller can read and write is called **low-level formatting,** or **physical formatting**.

4.3.1 Disk Formatting

- Low-level formatting fills the disk with a **special data structure** for each sector. The data structure for a sector typically consists of a **header, a data area** (usually **512 bytes in size**), and **a trailer.**
- The header and trailer contain information used by the disk controller, such as a **sector number and an error-correcting code (ECC).**
- When the **controller writes** a sector of data during normal I/O, the **ECC is updated** with a value calculated from all the bytes in the data area.
- When the **sector is read**, the **ECC is recalculated** and compared with he stored value.
- If the **stored and calculated numbers are different**, this mismatch indicates that the data area of the sector has become corrupted and that the disk sector may be bad.
- To use a disk to hold files, the operating system still needs to **record its own data structures on the disk.**
- It does so in **two steps**.
  - ➢ The **first step** - partition the disk into one or more groups of cylinders. The OS can treat each partition as a separate disk where one partition

can hold a copy of the OS's executable code, while another holds user files.

- ➢ The **second step** - logical formatting, or creation of a file system. The OS stores the initial file-system data structures onto the disk which may include maps of free and allocated space and an initial empty directory.

- **To increase efficiency**, most file systems group blocks together into larger chunks, frequently called **clusters**. Disk I/O is done via blocks, but file system I/O is done via clusters, effectively.

4.3.2 Boot Block

- For a computer to start running for instance, when it is powered up or rebooted—it must have an **initial program to run**. It initializes all aspects of the system, from CPU registers to device controllers and the contents of main memory, and then starts the operating system.
- To do its job, the bootstrap program **finds the OS kernel on disk**, loads that kernel into memory, and jumps to an initial address to begin the OS execution.
- For most computers, **the bootstrap is stored in read-only memory** (ROM). This location is convenient, because ROM needs no initialization and is at a fixed location that the processor can start executing when powered up or reset. And, since ROM is read only, it cannot be infected by a computer virus.
- The **problem** is that changing this bootstrap code requires changing the ROM hardware chips. For this reason, most systems store a tiny bootstrap loader program in the boot ROM whose only job is to bring in a full bootstrap program from disk.
- The full bootstrap program is stored in the "boot blocks" at a fixed location on the disk. A disk that has a boot partition is called a **boot disk or system disk.**
- The **code in the boot ROM** instructs the disk controller to **read the boot blocks into memory** and then **starts executing** that code. The full bootstrap program is more sophisticated than the bootstrap loader in the boot ROM. It is able to load the entire OS from a non-fixed location on disk and to start the OS running. Even so, the full bootstrap code may be small.
- **Example** the boot process in Windows.

First, Windows allows a hard disk to be divided into partitions, and one partition identified as the boot partition that contains the operating system and device drivers.

The Windows system places its boot code in the first sector on the hard disk, which it terms the master boot record, or MBR.

Booting begins by running code that is resident in the system's ROM memory. This code directs the system to read the boot code from the MBR. In addition to boot code, the MBR contains a table listing the partitions for the hard disk and a flag indicating which partition the system is to be booted from. Once the system identifies the boot partition, it reads the first sector from that partition (which is called **the boot sector**) and continues with the remainder of the boot process, which includes loading the various subsystems and system services.

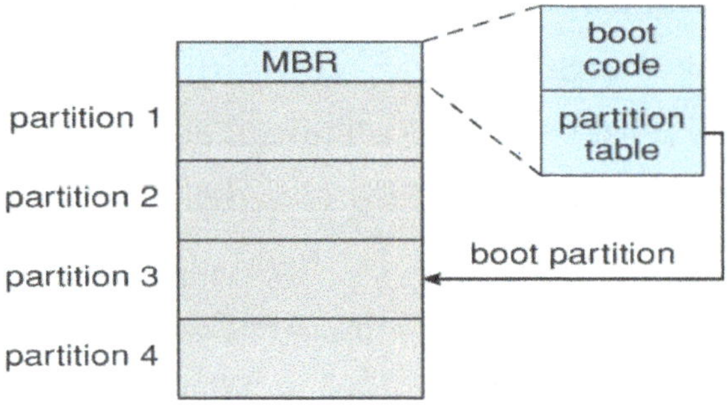

Figure 4.7 Boot block

4.3.3 Bad Blocks

- More sophisticated disks are smarter about bad-block recovery. The Controller maintains a list of **bad blocks on the disk**.
- The low-level formatting also sets aside **spare sectors** not visible to the OS. The controller can be told to replace each bad sector logically with one of the spare sectors. This scheme is known as **sector sparing or forwarding**.
- A typical **bad-sector transaction** might be **as follows**:
- The OS tries to read logical block 87. The controller calculates the ECC and finds that the sector is bad. It reports this finding to the OS and when the next time the system is rebooted, a special command is run to tell the controller to replace the bad sector with a spare.

- After that, whenever the system requests logical block 87, the request is translated into the replacement sector's address by the controller. As an alternative to **sector sparing**, some controllers can be instructed to replace a bad block by **sector slipping**.

Example

Suppose that logical block 17 becomes defective and the first available **spare follows sector 202**.

Sector slipping then remaps **all the sectors from 17 to 202**, moving them all down one spot. That is, sector 202 is copied into the spare, then sector 201 into 202, then 200 into 201, and so on, until sector 18 is copied into sector 19.

Slipping the sectors in this way frees up the space of sector 18 so that sector 17 can be mapped to it.

The replacement of a bad block generally is not totally automatic, because the data in the bad block are usually lost. Soft errors may trigger a process in which copy of the block data is made and the block is spared or slipped.

An unrecoverable **hard error** results in lost data and that requires manual intervention.

4.4 FILE SYSTEM STORAGE

Computers can **store information on various storage media**, such as magnetic disks, magnetic tapes, and optical disks. So that the **computer system will be convenient to use**, the **OS** provides **a uniform logical view of stored information**.

File Concepts

The OS abstracts from the physical properties of its storage devices to define a logical storage unit, the file. Files are mapped by the OS onto physical devices that are usually nonvolatile.

- A **file** is a named collection of related information that is recorded on secondary storage.

- From a user's perspective, a file is the smallest allotment of logical secondary storage ie, data cannot be written to secondary storage unless they are within a file.
- Commonly, files represent programs (both source and object forms) and data. Data files may be numeric, alphabetic, alphanumeric, or binary. Files may be free form, such as text files, or may be formatted rigidly.
- In general, a file is a sequence of bits, bytes, lines, or records.
- A file has a certain **defined structure**, which depends on its type.
    - A **text file** is a sequence of characters organized into lines (and possibly pages).
    - A **source file** is a sequence of functions, each of which is further organized as declarations followed by executable statements.
    - An **executable file** is a series of code sections that the loader can bring into memory and execute.

File Attributes

A file's attributes vary from one operating system to another but typically **consist of** these:

- **Name**. The symbolic file name is the only information kept in human readable form.
- **Identifier**. This unique tag, usually a number, identifies the file within the file system; it is the non-human-readable name for the file.
- **Type**. This information is needed for systems that support different types of files.
- **Location**. This information is a pointer to a device and to the location of the file on that device.
- **Size**. The current size of the file (in bytes, words, or blocks) and possibly the maximum allowed size are included in this attribute.
- **Protection**. Access-control information determines who can do reading, writing, executing, and so on.

- **Time, date, and user identification**. This information may be kept for creation, last modification, and last use. These data can be useful for protection, security, and usage monitoring

File Operations

A file is an abstract data type. The OS can provide system calls to create, write, read, reposition, delete, and truncate files.

- **Creating a file**. Two steps are necessary to create a file. First, space in the file system must be found for the file. Second, an entry for the new file must be made in the directory.
- **Writing a file**. To write a file, a system call is made specifying both the name of the file and the information to be written to the file. The system must keep a write pointer to the location in the file where the next write is to take place. The write pointer must be updated whenever a write occurs.
- **Reading a file**. To read from a file, a system call is used that specifies the name of the file and where (in memory) the next block of the file should be put. Again, the directory is searched for the associated entry, and the system needs to keep a read pointer to the location in the file where the next read is to take place. Both the read and write operations use this same pointer, saving space and reducing system complexity.
- **Repositioning within a file**. The directory is searched for the appropriate entry, and the current-file-position pointer is repositioned to a given value. This file operation is also known as a file seek.
- **Deleting a file**. To delete a file, the directory is searched for the named file. Having found the associated directory entry, all file space is released so that it can be reused by other files, and erase the directory entry.
- **Truncating a file**. The user may want to erase the contents of a file but keep its attributes. Rather than forcing the user to delete the file and then recreate it, this function allows all attributes to remain unchanged.

These six basic operations comprise the minimal set of required file operations.

- **Open(F_i)** – search the directory structure on disk for entry F_i, and move the content of entry to memory

- **Close (F_i)** – move the content of entry F_i in memory to directory structure on disk

There are several pieces of information that are associated with an open file.

- **File pointer**. On systems that do not include a file offset as part of the read () and write () system calls, the system must track the last read–write location as a current-file-position pointer.
- **File-open count.** The file-open count tracks the number of opens and closes and reaches zero on the last close. The system can then remove the entry.
- **Disk location of the file**. Most file operations require the system to modify data within the file. The information needed to locate the file on disk is kept in memory so that the system does not have to read it from disk for each operation.
- **Access rights**. Each process opens a file in an access mode. This information is stored on the per-process table so the OS can allow or deny subsequent I/O requests.

File locks provide **functionality similar** to **reader–writer locks**.

- A **shared lock** is akin to a reader lock in that several processes can acquire the lock concurrently.
- An **exclusive lock** behaves like a writer lock; only one process at a time can acquire such a lock.

OS may provide either mandatory or advisory **file-locking mechanisms**.

- If a lock is **mandatory**, then once a process acquires an exclusive lock, the operating system will prevent any other process from accessing the locked file.
- If the lock is **advisory**, then the operating system will not prevent the text editor from acquiring access

File Types

A file type includes the type as part of the file name. The name is split into two parts—a name and an extension, usually separated by a period. The user and the OS can tell from the name alone what the type of a file is.

Most OS allow users to specify a file name as a sequence of characters followed by a period and terminated by an extension made up of additional characters. Examples include resume.docx, server.c, and ReaderThread.cpp.

Table 4.1 File types

file type	usual extension	function
executable	exe, com, bin or none	ready-to-run machine-language program
object	obj, o	compiled, machine language, not linked
source code	c, cc, java, pas, asm, a	source code in various languages
batch	bat, sh	commands to the command interpreter
text	txt, doc	textual data, documents
word processor	wp, tex, rtf, doc	various word-processor formats
library	lib, a, so, dll	libraries of routines for programmers
print or view	ps, pdf, jpg	ASCII or binary file in a format for printing or viewing
archive	arc, zip, tar	related files grouped into one file, sometimes compressed, for archiving or storage
multimedia	mpeg, mov, rm, mp3, avi	binary file containing audio or A/V information

File Structure

File types also can be used to indicate the internal structure of the file. **Ex**. if the OS defines five different file structures, it needs to contain the code to support these file structures.

When new applications require information structured in ways not supported by the OS, severe problems may result.

For example, assume that a system supports two types of files:

- Text files and
- Executable binary files.

If (as users) want to **define an encrypted file** to protect the contents from being read by unauthorized people, **neither file type is found** to be appropriate.

The encrypted file is not ASCII text lines but rather is random bits.

Although it may appear to be a binary file, **it is not executable**.

Some OS support a **minimal number of file structures**. This approach has been adopted in UNIX, Windows, and others.

UNIX considers each file to be a sequence of 8-bit bytes; this **provides maximum flexibility** but little support.

Each application program must include its **own code to interpret an input file** as to the appropriate structure.

However, all **OS** must support **at least one structure**—that of an **executable file**—so that the system is able to load and run programs.

Internal File Structure

Disk systems have a **well-defined block size determined by the size of a sector.** All disk I/O is performed in units of one block (physical record), and all blocks are the same size.

The physical record size will exactly match the length of the desired logical record. Logical records may even vary in length. Packing a number of logical records into physical blocks is a **common solution to** this problem.

For example, the UNIX OS defines all files to be simply streams of bytes. Each byte is individually addressable by its offset from the beginning (or end) of the file.

In this case, the logical record size is 1 byte. The file system automatically packs and unpacks bytes into physical disk blocks— say, 512 bytes per block—as necessary.

The logical record size, physical block size, and packing technique determine **how many logical records are in each physical block.**

The packing can be done either by the user's application program or by the OS. In either case, the **file may be** considered **a sequence of blocks**.

The conversion from logical records to physical blocks is a relatively **simple software problem**.

Because disk space is always allocated in blocks, **some portion** of the last block of each file is **wasted.** If each block were 512 bytes, for example, then a file of 1,949 bytes would be allocated four blocks (2,048 bytes); the last 99 bytes would be wasted.

The waste incurred to keep everything in units of blocks (instead of bytes) is **internal fragmentation**. All file systems suffer from internal fragmentation; the larger the block size, the greater the internal fragmentation.

4.5 Directory and Disk Structure

File systems need to be organized because they store thousands, millions, even billions of files within a computer.

Files are stored on random-access storage devices, including hard disks, optical disks, and solid-state (memory-based) disks.

This organization is done in **two parts:**

- The disk can be partitioned into **one / more partitions**. Partitions also known as minidisks/ slices. Disk or partition can be used raw without a file system, or formatted with a file system.
- Any entity containing a file system is generally known as a **volume**. Each volume that contains a file system must also contain information about the files in the systems. This information is kept in entries in a **device directory or volume table of contents**. The device directory records information such as name, location, size, and type for all files on that volume.

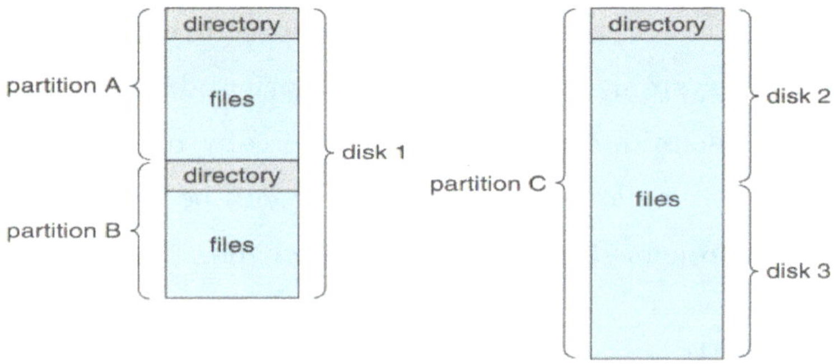

Figure 4.8 A typical file-system organization.

4.5.1 Storage Structure

Systems frequently have may file systems, some general- and some special-purpose.

Consider the types of file systems in the **Solaris example**

- tmpfs—a"temporary"file system that is created in volatile main memory and has its contents erased if the system reboots or crashes
- objfs—a"virtual"file system (looks like a file system) that gives debuggers access to kernel symbols
- ctfs—a virtual file system that maintains "contract" information to manage which processes start when the system boots and must continue to run during operation
- lofs—a"loop back"file system that allows one file system to be accessed in place of another one

- procfs—a virtual file system that presents information on all processes as a file system
- ufs, zfs—general-purpose file systems

Operations Performed on Directory

- Search for a file- Search a directory structure to find the entry for a particular file.
- Create a file- New files need to be created and added to the directory.
- Delete a file- When a file is no longer needed, we want to be able to remove it from the directory.
- List a directory- List the files in a directory and the contents of the directory entry for each file in the list.
- Rename a file- Allow its position within the directory structure to be changed.
- Traverse the file system. Access every directory and every file within a directory structure. This technique provides a backup copy in case of system failure. In addition, if a file is no longer in use, the file can be copied to tape and the disk space of that file released for reuse by another file.

4.5.2 Logical structure

4.5.2.1 Single-Level Directory

The simplest directory structure is the single-level directory. All files are contained in the same directory, which is easy to support and understand.

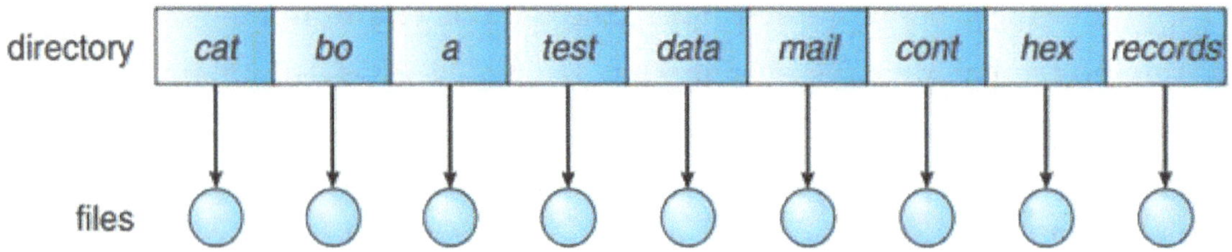

Figure 4.9 Single-Level Directory

A single-level directory has significant limitations, however, when the number of files increases or when the system has more than one user.

Since all files are in the same directory, they must have **unique names**. For example, in one programming class, 23 students called the program for their second assignment prog2.c; another 11 called it prog2.c, then unique rule is violated.

Fortunately, most file systems support file names of up to 255 characters, so it is relatively easy to select unique file names.

Even a single user on a single-level directory may find it difficult to remember the names of all the files as the number of files increases. So a two level directory is chosen.

4.5.2.2 Two-Level Directory

Problem in a single-level directory -> leads to **confusion of file names among different users**. Solution -> create a **separate directory for each user**.

In the two-level directory structure, each user has own **user file directory (UFD)**.

The UFDs have similar structures, but each lists only the files of a single user.

When a user job starts or a user logs in, the system's **master file directory** (MFD) is searched. The MFD is indexed by user name or account number, and each entry points to the UFD for that user.

When a user refers to a particular file, only his own UFD is searched. Thus, different users may have files with the same name, as long as all the file names within each UFD are unique.

To create a file for a user, the OS searches only that user's UFD to ascertain whether another file of that name exists. To delete a file, the OS confines its search to the local UFD; thus, it cannot accidentally delete another user's file that has the same name.

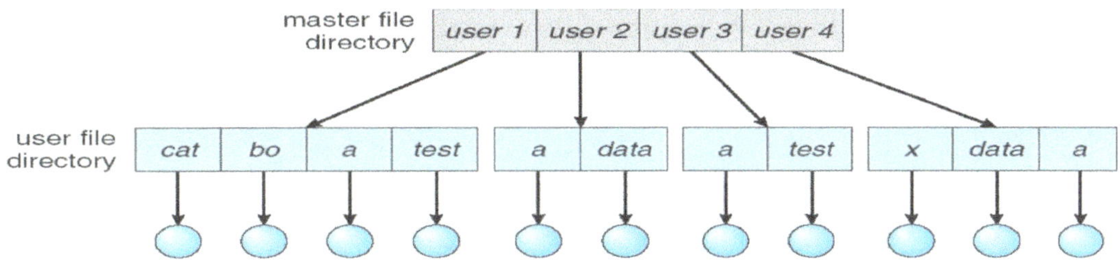

Figure 4.10 File-System Interface

A two-level directory can be thought of as a tree, or an inverted tree, of height 2.

The root of the tree is the MFD. Its direct descendants are the UFDs. The descendants of the UFDs are the files themselves.

The files are the leaves of the tree. Specifying a user name and a file name defines a path in the tree from the root (the MFD) to a leaf (the specified file).

Disadvantages of Two-Level Directory:

It effectively isolates one user from another- when the users want to cooperate on some task and to access one another's files.

Some systems simply do not allow local user files to be accessed by other users- If access is to be permitted, one user must have the ability to name a file in another user's directory.

4.5.2.3 Tree-Structured Directories

To view a two-level directory as a two-level tree, the natural generalization is to extend the directory structure to a tree of arbitrary height. This generalization allows users to create their own subdirectories and to organize their files accordingly.

A **tree** is the most common directory structure. The tree has a root directory, and every file in the system has a unique path name.

A **directory** (or subdirectory) contains a set of files or subdirectories. A directory is simply another file, but it is treated in a special way.

All directories have the same internal format. One bit in each directory entry defines the entry as a file (0) or as a subdirectory (1). Special system calls are used to create and delete directories.

Path names can be of **two types**: absolute and relative.

An absolute path name begins at the root and follows a path down to the specified file, giving the directory names on the path.

A relative path name defines a path from the current directory.

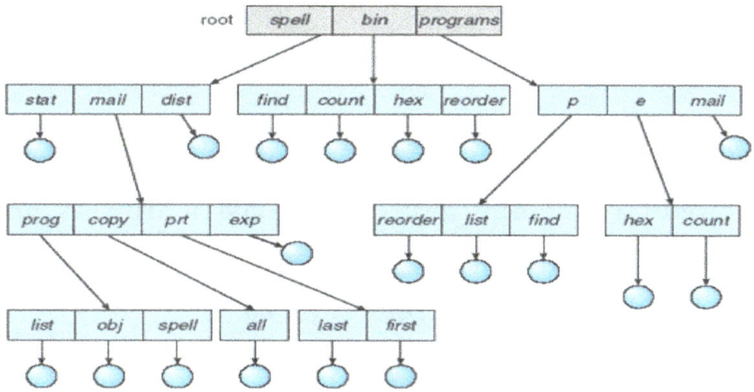

Figure 4.11 Tree-structured directory structure

For example, in the tree-structured file system, if the current directory is root/spell/mail, then the **relative path name** prt/first refers to the same file as does the **absolute path name** root/spell/mail/prt/first.

With a tree-structured directory system, **users can be allowed to access, in addition to their files, the files of other users.**

For example, user B can access a file of user A by specifying its path names. User B can specify either an absolute or a relative path name. Alternatively, user B can change her current directory to be user A's directory and access the file by its file names.

4.5.2.4 Acyclic-Graph Directories

A tree structure prohibits the **sharing of files or directories**.

An acyclic graph is a graph with no cycles - allows directories to share subdirectories and files.

The acyclic graph is a natural generalization of the tree-structured directory scheme.

With a shared file,

- It is not the same as two copies of the file. With two copies, each programmer can view the copy rather than the original, but if one programmer changes the file, the changes will not appear in the other's copy.
- Only one actual file exists, so any changes made by one person are immediately visible to the other.

- A new file created by one person will automatically appear in all the shared subdirectories.

Shared files and subdirectories **can be implemented in several ways**.

1) one way is to create a new directory entry called **a link**. A link is effectively a pointer to another file or subdirectory.

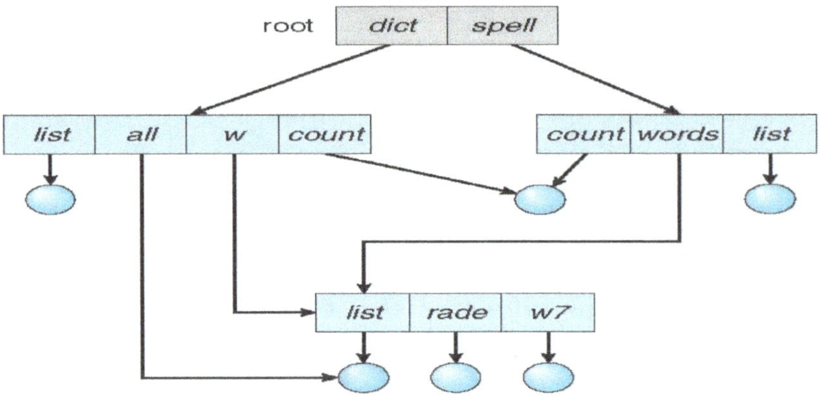

Figure 4.12 Acyclic-graph directory structure.

For **example,** a link may be implemented as an absolute or a relative path name.

When a reference to a file is made, the directory is searched. If the directory entry is marked as a link, then the name of the real file is included in the link information.

The link is resolved by using that path name to locate the real file. Links are easily identified by their format in the directory entry and are effectively indirect pointers.

The OS ignores these links when traversing directory trees to preserve the acyclic structure of the system.

2) Another is to duplicate all information about them in both sharing directories. Thus, both entries are identical and equal.

Difference between this approach and the creation of a link.

- The link is clearly different from the original directory entry; thus, the two are not equal.
- Duplicate directory entries make the original and the copy indistinguishable. Problem : maintaining consistency when a file is modified.

Advantage

An acyclic-graph directory structure is **more flexible** than a simple tree structure, but it is also more complex.

Several problems must be considered.

- **Aliasing problem**

A file may now have multiple absolute path names. The distinct file names may refer to the same file. This situation is similar to the **aliasing problem** for programming languages.

- **Deletion**

The space allocated to a shared file can be deallocated and reused whenever any of the file is deleted, but this action may leave **dangling pointers** to the nonexistent file.

Approaches:

1) When sharing is implemented by **symbolic links**, the deletion of a link need not affect the original file; only the link is removed. If the file entry itself is deleted, the space for the file is deallocated, leaving the links dangling.
2) Preserve the file until all references to it are deleted. The trouble with this approach
   - Variable and
   - Large size of the file-reference list.

   It need not have to keep the entire list, but is required to keep only a count of the number of references.
   - Adding a new link or directory entry increments the reference count.
   - Deleting a link or entry decrements the count. When the count is 0, the file can be deleted; there are no remaining references to it.

4.5.2.5 General Graph Directory

Problem with using an acyclic-graph: ensuring that there are no cycles.

If a **two-level directory** is started it allows users to create subdirectories, a **tree-structured directory results**. It should be fairly easy to see that simply adding new files and subdirectories to an existing tree-structured directory preserves the tree-structured nature.

But **when added links**, the tree **structure is destroyed**, **resulting in a simple graph structure**.

Advantage of an acyclic graph

- relative simplicity of the algorithms to traverse the graph and
- determine when there are no more references to a file.

The traversing of shared sections of an acyclic graph must be avoided twice, mainly for performance reasons.

- If a major shared subdirectory is searched for a particular file without finding it, then searching that subdirectory is avoided again; the second search would be a waste of time.
- If cycles are allowed to exist in the directory, then avoid searching any component twice, for reasons of correctness as well as performance.

An algorithm might result in an infinite loop continually searching through the cycle and never terminating.

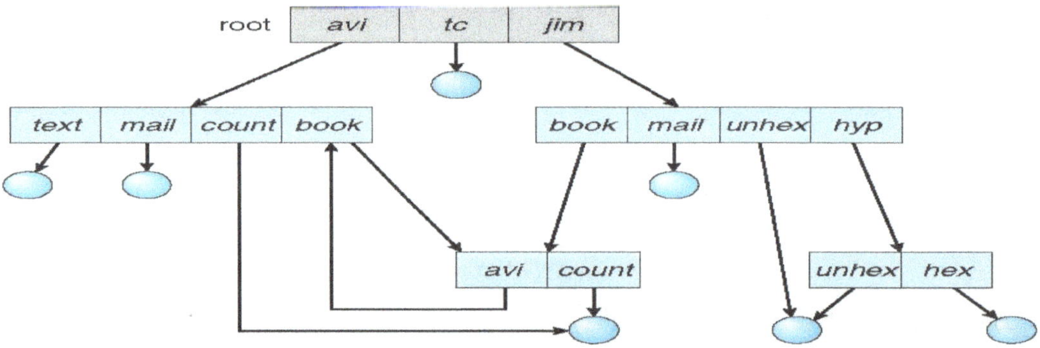

Figure 4.13 General graph directory.

One solution is to limit arbitrarily the number of directories that will be accessed during a search.

A similar problem exists when trying to determine when a file can be deleted. With acyclic-graph directory structures, a value of 0 in the reference count means that there are no more references to the file or directory, and the file can be deleted.

4.6 File Sharing

-Sharing of files on multi-user systems is desirable.

-Sharing may be done through a protection scheme.

-On distributed systems, files may be shared across a network.

-Network File System (NFS) is a common distributed file-sharing method.

4.6.1 Multiple Users

On a multi-user system, more information needs to be stored for each file:

- **User IDs** - The owner (user) who owns the file can control its access.
- **Group IDs** - The group of other user IDs that may have some special access to the file.
- The access rights are afforded to the owner, the Group, and to the rest of the world.
- Some systems have more complicated access control, allowing or denying specific accesses to specifically named users.

4.6.2 Remote File Systems

The advent of the Internet introduces issues for accessing files stored on remote computers

- **Manually through programs like FTP**, allowing individual files to be transported across systems as needed. Ftp can be either account or password controlled, or anonymous, not requiring any user name or password.
- **Automatically & seamlessly using distributed file systems** allow remote file systems to be mounted onto a local directory structure, and accessed using normal file access commands.
- **Semi-automatically via WWW** has made it easy once again to access files on remote systems without mounting their file systems, generally using ftp as the underlying file transport mechanism.

a) The Client –Server Model

- When one computer system remotely mounts a file system that is physically located on another system, the system which physically owns the files acts as a **server**, and the system which mounts them is the **client.**

- User IDs and group IDs must be consistent across both systems for the system to work properly.
- The same computer can be both a client and a server. (E.g. cross-linked file systems.)
- There are a number of security concerns involved in this model:
    - Servers commonly restrict mount permission to certain trusted systems only. Spoofing (a computer pretending to be a different computer) is a potential security risk.
    - Servers may restrict remote access to read-only.
    - Servers restrict which file systems may be remotely mounted. Generally the information within those subsystems is limited, relatively public, and protected by frequent backups.
- The NFS (Network File System) is a classic example of such a system.

b) Distributed Information Systems

- The **Domain Name System, DNS,** provides for a unique naming system across all of the Internet.
- Domain names are maintained by the **Network Information System, NIS**, which unfortunately has several security issues. NIS+ is a more secure version, but has not yet gained the same widespread acceptance as NIS.
- Microsoft's **Common Internet File System, CIFS**, establishes a **network login** for each user on a networked system with shared file access. Older Windows systems used **domains**, and newer systems (XP, 2000), use **active directories.** User names must match across the network for this system to be valid.
- A newer approach is the **Lightweight Directory-Access Protocol, LDAP,** which provides a **secure single sign-on** for all users to access all resources on a network. This is a secure system and has the maintenance advantage of combining authorization information in one central location.

c) Failure Modes

- All file systems have failure modes. For example corruption of directory structures or other non-user data, called metadata

- When a local disk file is unavailable, the result is generally known immediately, and is generally non-recoverable. The only reasonable response is for the response to fail.
- Remote file systems add new failure modes, due to network failure & server failure
- Recovery from failure can involve state information about status of each remote request
- Stateless protocols such as NFS v3 include all information in each request, allowing easy recovery but less security

4.6.3 Consistency Semantics

- Consistency Semantics deals with how multiple users of a system are to access a shared file simultaneously. ie. When one user changes the file, when do other users see the changes?
- Consistency semantics are directly related to the process synchronization algorithms.

a) UNIX Semantics

The UNIX file system uses the following consistency semantics:

- Writes to an open file by a user are visible immediately to other users who have this file open.
- One mode of sharing allows users to share the pointer of current location into the file. Thus, the advancing of the pointer by one user affects all sharing users. Here, a file has a single image that interleaves all accesses, regardless of their origin.
- In the UNIX semantics, a file is associated with a single physical image that is accessed as an exclusive resource. Contention for this single image causes delays in user processes.

b) Session Semantics

The Andrew file system (Open AFS) uses the following consistency semantics:

- Writes to an open file by a user are not visible immediately to other users that have the same file open.
- Once a file is closed, the changes made to it are visible only in sessions starting later. Already open instances of the file do not reflect these changes.

- According to these semantics, a file may be associated temporarily with several images at the same time. The multiple users are allowed to perform both read and write accesses concurrently on their images of the file, without delay.

c) Immutable-Shared-Files Semantics

- A unique approach is that of immutable shared files. Once a file is declared as shared by its creator, it cannot be modified.
- An immutable file has **two key properties**:
- its name may not be reused, and
- its contents may not be altered.
- Thus, the name of an immutable file signifies that the contents of the file are fixed.
- The implementation of these semantics in a distributed system is simple, because the sharing is disciplined (read-only).

4.7 PROTECTION

- When information is stored in a computer system, it is **required to keep safe** from physical damage (the issue of reliability) and improper access (the issue of protection).
- **Reliability** is generally provided by duplicate copies of files.
- File systems can be **damaged by** hardware problems (such as errors in reading or writing), power surges or failures, head crashes, dirt, temperature extremes, and vandalism.
- Files may be **deleted accidentally**. Bugs in the file-system software can also cause file contents to be lost.
- **Protection can be provided** in many ways. Ex.: For a single-user laptop system, protection is provided by locking the computer in a desk drawer or file cabinet.

Types of Access

The need to protect files is a direct result of the ability to access files. Systems that do not permit access to the files of other users do not need protection.

Protection mechanisms provide controlled access by limiting the types of file access that can be made.

Access is permitted or denied depending on several factors, one of which is the type of access requested.

Several **different types of operations** may be controlled:

- **Read**. Read from the file.
- **Write**. Write or rewrite the file.
- **Execute**. Load the file into memory and execute it.
- **Append**. Write new information at the end of the file.
- **Delete**. Delete the file and free its space for possible reuse.
- **List**. List the name and attributes of the file.

Other operations, such as **renaming, copying, and editing the file**, may also be controlled.

Access Control

The most common approach to the protection problem is to make access dependent on the identity of the user. Different users may need different types of access to a file or directory.

The most general scheme to implement identity dependent access is to associate with each file and directory an access-control list (ACL) specifying user names and the types of access allowed for each user.

When a user requests access to a particular file, the operating system checks the access list associated with that file.

If that user is listed for the requested access, the access is allowed. Otherwise, a protection violation occurs, and the user job is denied access to the file.

Advantage- complex access methodologies.

Problem- length. If everyone are allowed to read a file, all users are listed with read access.

This technique has **two undesirable consequences**:

- It may be a tedious if the **list of users** in the system is **not know in advance**.
- The directory entry, previously of fixed size, now must be of variable size, resulting in more **complicated space management**.

- These **problems can be resolved** by use of a condensed version of the access list. There are **three classifications** of users in connection with each file:
- **Owner**. The user who created the file is the owner.
- **Group.** A set of users who are sharing the file and need similar access is a group, or work group.
- **Universe**. All other users in the system constitute the universe.

Each field is a collection of bits, and each bit either allows or prevents the access associated with it.

For example, the UNIX system **defines three fields** of 3 bits each—**rwx**, where

- r controls read access,
- w controls write access, and
- x controls execution.

A separate field is kept for the file owner, for the file's group, and for all other users.

In this scheme, 9 bits per file are needed to record protection information.

4.8 FILE-SYSTEM IMPLEMENTATION

4.8.1 File System Structure

Disks provide most of the secondary storage on which file systems are maintained. **Two characteristics** make them convenient for this purpose:

1. A disk can be rewritten in place; it is possible to read a block from the disk, modify the block, and write it back into the same place.
2. A disk can access directly any block of information it contains. Thus, it is simple to access any file either sequentially or randomly, and switching from one file to another requires only moving the read–write heads and waiting for the disk to rotate.

To improve I/O efficiency, I/O transfers between memory and disk are performed in units of blocks. Each block has one or more sectors. Depending on the disk drive, sector size varies from 32 bytes to 4,096 bytes; the usual size is 512 bytes.

File systems provide efficient and convenient access to the disk by allowing data to be stored, located, and retrieved easily.

Problems:

1) Defining how the file system should look to the user - involves defining a file and its attributes, the operations allowed on a file, and the directory structure for organizing files.
2) Creating algorithms and data structures to map the logical file system onto the physical secondary-storage devices.

The file system itself is composed of many different levels. Each level in the design uses the features of lower levels to create new features for use by higher levels.

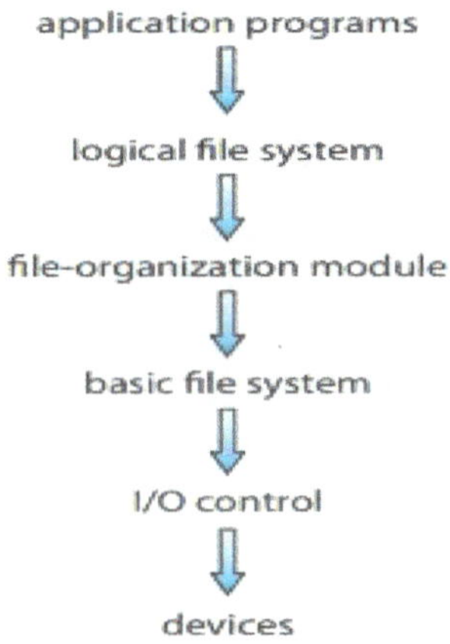

Figure 4.14 Layered File System

1) **The I/O control** level consists of device drivers and interrupts handlers to transfer information between the main memory and the disk system.

A device driver

- is a translator.
- its input consists of high level commands.
- its output consists of low-level, hardware-specific instructions that are used by the hardware controller, which interfaces the I/O device to the rest of the system.
- it usually writes specific bit patterns to special locations in the I/O controller's memory.

2) **The basic file system** needs only to issue generic commands to the appropriate device driver- to read and write physical blocks on the disk.

   This layer also manages the memory buffers and caches that hold various file-system, directory, and data blocks.

3) **The file-organization** module knows about files and their logical blocks, as well as physical blocks.

It can translate logical block addresses to physical block addresses for the basic file system to transfer. The file-organization module also includes the free-space manager, which tracks unallocated blocks and provides these blocks to the file-organization module when requested.

4) **The logical file system** manages metadata information. Metadata includes all of the file-system structure except the actual data (or contents of the files).

It manages the directory structure to provide the file-organization module with the information the latter needs, given a symbolic file name.

It maintains file structure via file-control blocks.

5) **A file control block** (FCB) contains information about the file, including ownership, permissions, and location of the file contents. The logical file system is also responsible for protection.

4.8.2 File-System Implementation -Overview

Several on-disk and in-memory structures are used to implement a file system. These structures vary depending on the operating system and the file system.

On disk, the file system may contain information about how to boot an operating system stored there, the total number of blocks, the number and location of free blocks, the directory structure, and individual files.

•**A boot control block** (per volume) can contain information needed by the system to boot an operating system from that volume. If the disk does not contain an operating system, this block can be empty. It is typically the first block of a volume.

- In UFS, it is called the boot block.
- In NTFS, it is the partition boot sector.

•**A volume control block** (per volume) contains volume (or partition) details, such as the number of blocks in the partition, the size of the blocks, a free-block count and free-block pointers, and a free-FCB count and FCB pointers.

- In UFS, this is called a superblock.

- In NTFS, it is stored in the master file table.

• **A directory structure** (per file system) is used to organize the files.

- In UFS, this includes file names and associated inode numbers.
- In NTFS, it is stored in the master file table.

• **A per-file FCB** contains many details about the file. It has a unique identifier number to allow association with a directory entry.

- In NTFS, this information is actually stored within the master file table, which uses a relational database structure, with a row per file

The in-memory information is used for both file-system management and performance improvement via caching. The data are loaded at mount time, updated during file-system operations, and discarded at dismount.

Several types of structures may be included.

- An in-memory **mount table** contains information about each mounted volume.
- An in-memory **directory-structure cache** holds the directory information of recently accessed directories. (For directories at which volumes are mounted, it can contain a pointer to the volume table.)
- The **system-wide open-file table** contains a copy of the FCB of each open file, as well as other information.
- The **per-process open-file table** contains a pointer to the appropriate entry in the system-wide open-file table, as well as other information.
- Buffers hold **file-system blocks** when they are being read from disk or written to disk.

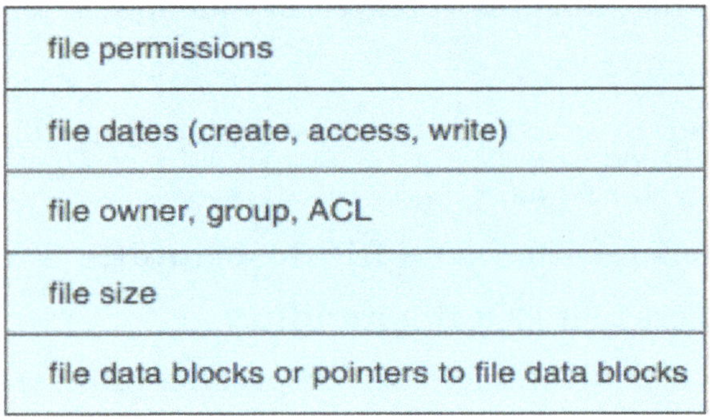

Figure 4.15 A Typical File-Control Block

To create a new file, an application program calls the logical file system. The logical file system knows the format of the directory structures. To create a new file, it allocates a new FCB.

The operating structures of a file-system implementation:

- A file has been created; it can be used for I/O. First, it must be opened. The open () call passes a file name to the logical file system.
- The open () system call first searches the system-wide open-file table to see if the file is already in use by another process.
- If it is, a per-process open-file table entry is created pointing to the existing system-wide open-file table.
- If the file is not already open, the directory structure is searched for the given file name.
- Once the file is found, the FCB is copied into a system-wide open-file table in memory. This table not only stores the FCB but also tracks the number of processes that have the file open.
- Next, an entry is made in the per-process open-file table, with a pointer to the entry in the system-wide open-file table and some other fields.
- These may include a pointer to the current location in the file (for the next read () or write () operation) and the access mode in which the file is open.
- The open () call returns a pointer to the appropriate entry in the per-process.
- All file operations are then performed via this pointer. The file name may not be part of the open-file table, as the system has no use for it once the appropriate FCB is located on disk.
- It could be cached, to save time on subsequent opens of the same file. The name given to the entry varies as,
    - UNIX systems refer to it as a **file descriptor**;
    - Windows refers to it as **a file handle**.
- When a process closes the file, the per-process table entry is removed, and the system-wide entry's open count is decremented.

- When all users that have opened the file close it, any updated metadata is copied back to the disk-based directory structure, and the system-wide open-file table entry is removed.

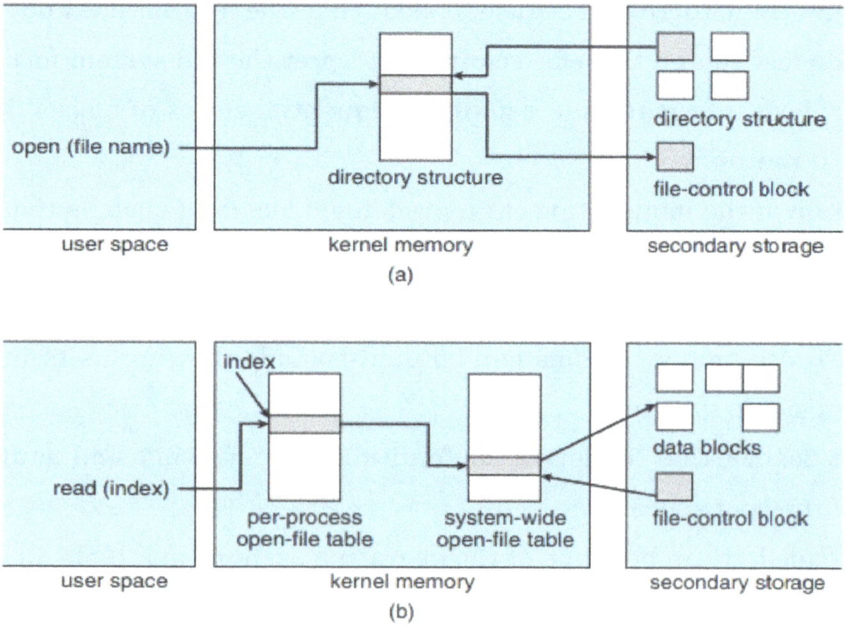

Figure 4.16 In- memory file-system structures a) File Open b) File Read

4.8.3 Partitions and Mounting

The layout of a disk can have many variations, depending on the OS. A disk can be sliced into multiple partitions, or a volume can span multiple partitions on multiple disks.

Considering the first category:

- Each partition can be either **"raw,"** containing no file system, **or "cooked,"** containing a file system.

  Raw disk is used where no file system is appropriate. UNIX swap space can use a raw partition, for example, since it uses its own format on disk and does not use a file system.

  Raw disk can also hold information needed by disk RAID systems, such as bit maps indicating which blocks are mirrored and which have changed and need to be mirrored.

Raw disk can contain a miniature database holding RAID configuration information, such as which disks are members of each RAID set.

➢ **Boot information** can be stored in a separate partition.

It has its **own format**, because at boot time the system does not have the file-system code loaded and therefore cannot interpret the file-system format.

Rather, boot information is usually a sequential series of blocks, **loaded as an image into memory.**

- Execution of the image starts at a predefined location, such as the first byte.
- This boot loader in turn knows enough about the file-system structure to load the kernel and start it executing.
- For instance, many systems can be dual-booted, allowing us to install multiple OS on a single system.
- A boot loader that understands multiple file systems and multiple OS can occupy the boot space.
- Once loaded, it can boot one of the operating systems available on the disk.
- The disk can have multiple partitions, each containing a different type of file system and a different OS.

➢ **The root partition**, which contains the operating-system kernel and sometimes other system files, is mounted at boot time.

Other volumes can be automatically mounted at boot or manually mounted later, depending on the OS.

The OS verifies with the device driver that the device contains a valid file system or not.

If the format is invalid, the partition must have its consistency checked and possibly corrected, either with or without user intervention.

Finally, the OS notes in its in-memory mount table that a file system is mounted, along with the type of the file system.

➢ **Microsoft Windows**–based systems mount each volume in a separate name space, denoted by a letter and a colon. To record that a file system is mounted at F.

On UNIX, file systems can be mounted at any directory. Mounting is implemented by setting a flag in the in-memory copy of the inode for that directory. The flag indicates that the directory is a mount point.

4.8.3.1 Virtual File Systems

The file-system implementation consists of **three major layers**

- The **first layer** is the **file-system interface**, based on the open(), read(),write(),and close() calls and on file descriptors.
- The **second layer** is called the **virtual file system (VFS) layer**.

The VFS layer serves **two important functions**:

1. It separates file-system-generic operations from their implementation by defining a clean VFS interface.
2. The VFS is based on a file-representation structure, called a vnode that contains a numerical designator for support of network file systems. (UNIX inodes are unique within only a single file system.) The kernel maintains one vnode structure for each active node (file or directory).

Thus, the VFS distinguishes local files from remote ones, and local files are further distinguished according to their file-system types.

The VFS activates file-system-specific operations to handle local requests according to their file-system types and calls the NFS protocol procedures for remote requests.

- The layer implementing the **file-system type or the remote-file-system** protocol is the **third layer of the architecture.**

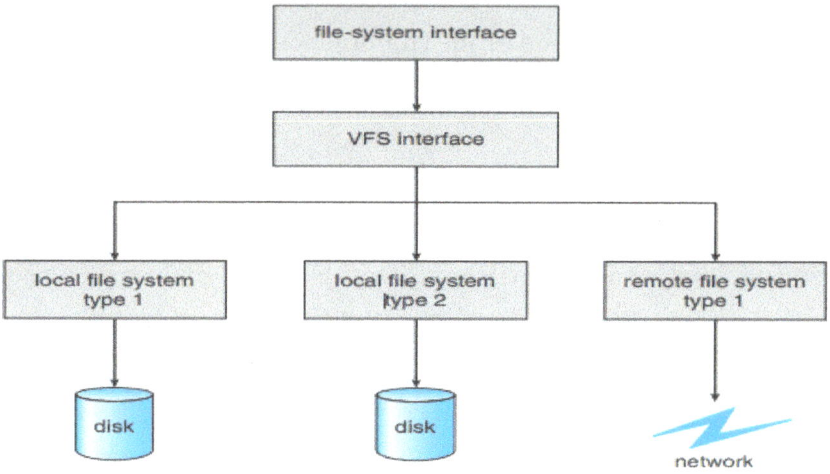

Figure 4.17 Schematic View of a Virtual File System

VFS architecture in Linux:

The four main object types defined by the Linux VFS are:

- The **inode object**, which represents an individual file
- The **file object**, which represents an open file
- The **superblock object**, which represents an entire file system
- The **dentry object**, which represents an individual directory entry

For each of these four object types, the VFS defines a set of operations that may be implemented. Every object of one of these types contains a pointer to a function table.

The function table lists the addresses of the actual functions that implement the defined operations for that particular object.

For example, an abbreviated API for some of the operations for the file object includes:

- int open(. . .)—Open a file.
- int close(...)—Close an already-open file.
- ssizet read(. . .)—Read from a file.
- ssizet write(. . .)—Write to a file.
- int mmap(. . .)—Memory-map a file.

Thus, the VFS software layer can perform an operation on one of these objects by calling the appropriate function from the object's function table, without having to know in advance exactly what kind of object it is dealing with.

4.9 Allocation Methods

The files are stored on the same disk. The main problem is how to allocate space to these files so that disk space is utilized effectively and files can be accessed quickly. There are three major methods of allocating disk space are in wide use:

- contiguous,
- linked, and
- indexed.

Each method has advantages and disadvantages.

4.9.1 Contiguous Allocation

Contiguous allocation requires that each file occupy a set of contiguous blocks on the disk. Disk addresses define a linear ordering on the disk.

With this ordering, assuming that only one job is accessing the disk, accessing block b+ 1after block b normally requires no head movement.

When head movement is needed (from the last sector of one cylinder to the first sector of the next cylinder), the head need only move from one track to the next.

Thus, the number of disk seeks required for accessing contiguously allocated files is minimal, as is seek time when a seek is finally needed.

Accessing a file that has been allocated contiguously is easy.

- For **sequential access**, the file system remembers the disk address of the last block referenced and, when necessary, reads the next block.
- For **direct access** to block i of a file that starts at block b, we can immediately access block b+i.

Thus, both sequential and direct access can be supported by contiguous allocation.

Drawback:

1) Finding space for a new file.

If the file to be created is n blocks long, then the OS must search for n free contiguous blocks.

First-fit, best-fit, and worst-fit strategies are the most common strategies used to select a free hole from the set of available holes.

Simulations have shown that both first-fit and best-fit are better than worst-fit in terms of both time storage utilization.

Neither first-fit nor best-fit is clearly best in terms of storage utilization, but first-fit is generally faster.

All these algorithms suffer from the problem of **external fragmentation**.

As files are allocated and deleted, the free disk space is broken into little pieces. External fragmentation exists when enough total disk space exists to satisfy a request, but this space not contiguous; storage is fragmented into a large number of small holes.

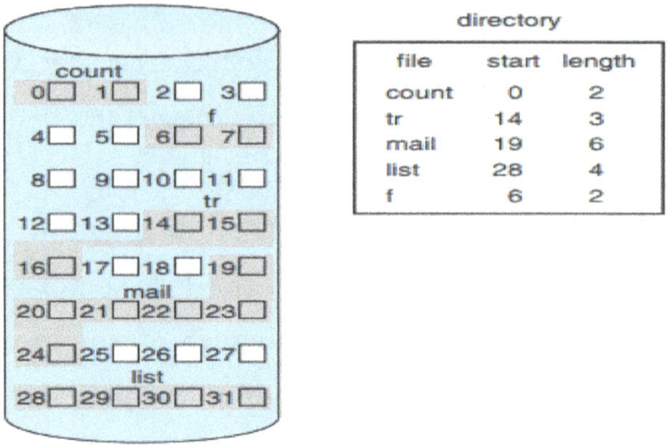

Figure 4.18 Contiguous allocation of disk space.

2) How much disk space is needed for a file.

When the file is created, the total amount of space it will need must be known and allocated. The size of the file to be created is not known. In some cases, this determination may be fairly simple (e.g. copying an existing file), but in general the size of an output file may be difficult to estimate.

4.9.2 Linked Allocation

Linked allocation solves all problems of contiguous allocation. With linked allocation, each file is a linked list of disk blocks; the disk blocks may be scattered anywhere on the disk. The directory contains a pointer to the first and last blocks of the file.

For example, a file of five blocks might start at block 9 and continue at block 16, then block 1, then block 10, and finally block 25. Each block contains a pointer to the next block. These pointers are not made available to the user. Thus, if each block is 512 bytes in size, and a disk address (the pointer) requires 4 bytes, then the user sees blocks of 508 bytes.

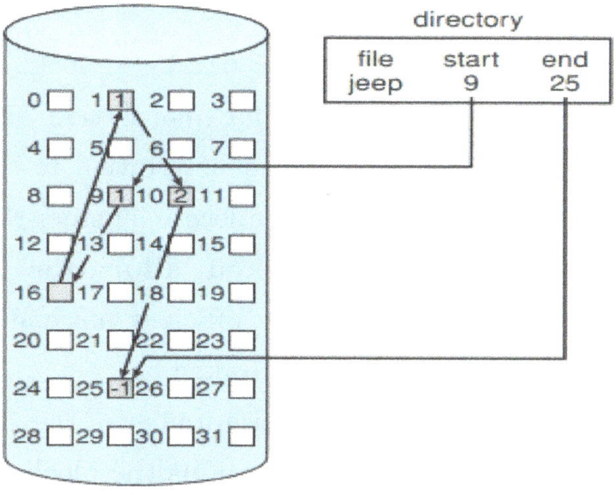

Figure 4.19 Linked allocation of disk space

With linked allocation, each directory entry has a pointer to the first disk block of the file. This pointer is initialized to nil (the end-of-list pointer value) to signify an empty file.

A write to a file removes the first free block and writes to that block. This new block is then linked to the end of the file. To read a file, the pointers are just followed from block to block.

There is **no external fragmentation** with linked allocation. Any free block can be used to satisfy a request. There is no need to declare the size of a file when that file is created. A file can continue to grow as long as there are free blocks.

Disadvantages.

1) Inefficient to support direct-access

Iit is effective only for sequential-access files. To find the ith block of a file, it must start at the beginning of that file and follow the pointers until the ith block is reached. Each access to a pointer requires a disk read.

2) Reliability

A bug in OS or disk hardware failure might result in pointers being lost and damaged. The effect of which could be picking up a wrong pointer and linking it to a free block or into another file.

3) The space required for the pointers

If a pointer requires 4 bytes out of a 512-byte block, then 0.78 percent of the disk is being used for pointers, rather than for information. Each file requires slightly more space than it would otherwise. The usual solution to this problem is to collect blocks into multiples, called clusters, and to allocate clusters rather than blocks.

An important variation on linked allocation is the **use of a file-allocation table (FAT)**. This is simple but efficient method of disk-space allocation. The FAT is used in much the same way as a linked list.

The directory entry contains the block number of the first block of the file. The table entry indexed by that block number contains the block number of the next block in the file. This chain continues until it reaches the last block.

A benefit is that random-access time is improved by reading the information in the FAT.

4.9.3 Indexed Allocation

Linked allocation solves the external-fragmentation and size-declaration problems of contiguous allocation.

But in the absence of a FAT, **linked allocation cannot support efficient direct access**, since the pointers to the blocks are scattered with the blocks themselves all over the disk and must be retrieved in order.

Indexed allocation solves this problem by bringing all the pointers together into one location: the **index block**.

Each file has its own index block, which is an array of disk-block addresses. The ith entry in the index block points to the ith block of the file. The directory contains the address of the index block.

To read the ith sector of the file, the pointer in the ith index block entry is read to find the desired sector.

Indexed allocation supports direct access, without suffering from external fragmentation. Any free block anywhere on the disk may satisfy a request for more space.

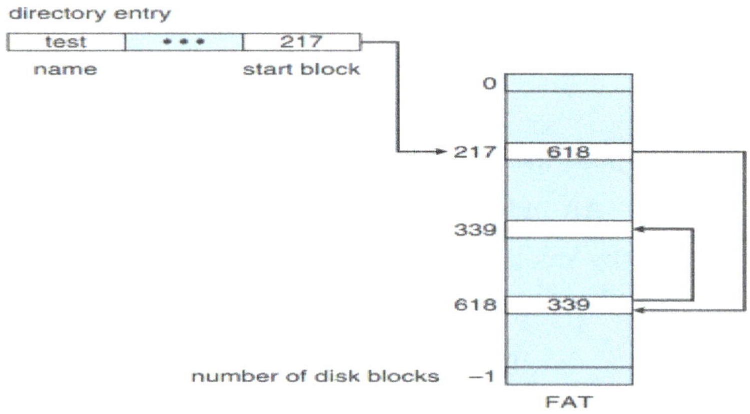

Figure 4.20 File-allocation table

Indexed allocation **supports direct access**, **without** suffering from **external fragmentation**, because any free block on the disk can satisfy a request for more space.

Indexed allocation does suffer from wasted space, however.

The pointer directory overhead of the index block is **greater** than the pointer overhead of linked allocation.

Example: Consider a file of only one or two blocks.

- With linked allocation, the space of only one pointer per block is lost.
- With indexed allocation, an entire index block must be allocated, even if only one or two pointers will be non-null.

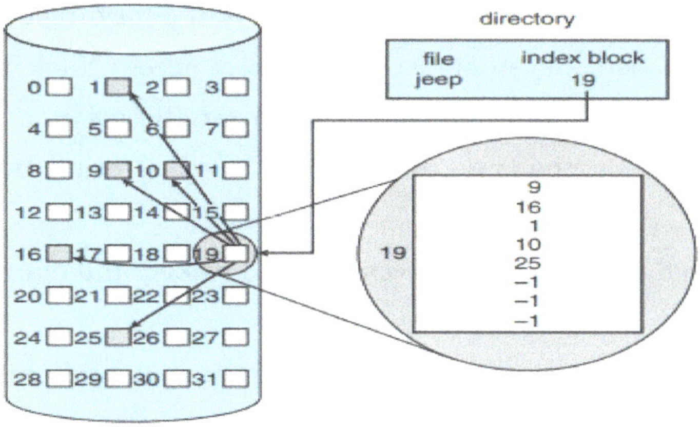

Figure 4.21 Indexed allocation of disk space

If the index block is too small, it will not be able to hold enough pointers for a large file, and a mechanism will have to be available to deal with this issue.

Mechanisms for this purpose include the following:

•**Linked scheme**- An index block is normally one disk block. Thus, it can be read and written directly by itself. To allow for large files, several index blocks can be linked.

•**Multilevel index**- A variant of linked representation uses a first-level index block to point to a set of second-level index blocks, which in turn point to the file blocks. To access a block, the OS uses the first-level index to find a second-level index block and then uses that block to find the desired data block. This approach could be continued to a third or fourth level, depending on the desired maximum file size.

Performance

The **allocation methods vary in their storage efficiency** and **data-block access times**.

Contiguous allocation, a system with mostly sequential access should not use the same method as a system with mostly random access. For any type of access, requires only one access to get a disk block. Since the initial address of the file is kept in memory, the disk address of the ith block can be calculated and read it directly.

Linked allocation, the address of the next block in memory is kept and read it directly. This method is fine for sequential access & not for direct access.

As a result, some systems support direct-access files by using contiguous allocation and sequential-access files by using linked allocation.

For these systems, the **type of access** to be made **must be declared** when the file is created.

- A file created for sequential access will be linked and cannot be used for direct access.
- A file created for direct access will be contiguous and can support both direct access and sequential access, but its maximum length must be declared when it is created.

Indexed allocation, is more complex. If the index block is already in memory, then the access can be made directly.

For a two-level index, two index-block reads might be necessary. For an extremely large file, accessing a block near the end of the file would require reading in all the index blocks before the needed data block finally could be read.

Thus, the **performance** of indexed allocation **depends on the index structure, on the size of the file, and on the position of the block desired**.

Some systems **combine contiguous allocation with indexed allocation** by using contiguous allocation for small files (up to three or four blocks) and automatically switching to an indexed allocation if the file grows large.

Since most files are small, and contiguous allocation is efficient for small files, **average performance** can be quite **good.**

4.10 Free Space Management

Since there is only a limited amount of disk space, it is necessary to reuse the space from deleted files for new files. To keep track of free disk space, the system maintains a **free-space list**.

The free-space list records all disk blocks that are free (i.e., are not allocated to some file).

To create a file, the free-space list has to be searched for the required amount of space, and allocate that space to a new file. This space is then removed from the free-space list.

When a file is deleted, its disk space is added to the free-space list.

4.10.1 Bit-Vector

Frequently, the free-space list is implemented as a **bit map or bit vector**. Each block is represented by a 1 bit. If the block is free, the bit is 0; if the block is allocated, the bit is 1.

For example, consider a disk where blocks 2, 3, 4, 5, 8, 9, 10, 11, 12, 13, 17, 18, 25, 26, and 27 are free, and the rest of the blocks are allocated. The free-space bit map would be:

11000011000000111001111110001111...

Advantage:

- relatively simple
- efficient (to find n consecutive free blocks on the disk.)

The bit vectors are inefficient unless the entire vector is kept in memory for most accesses. Keeping it main memory is possible for smaller disks such as on microcomputers, but not for larger ones.

4.10.2 Linked List

Another approach is to link all the free disk blocks together, keeping a pointer to the first free block. This block contains a pointer to the next free disk block, and so on.

In the previous example, a pointer could be kept to block 2, as the first free block. Block 2 would contain a pointer to block 3, which would point to block 4, which would point to block 5, which would point to block 8, and so on.

This scheme is not efficient; to traverse the list, each block must be read, which requires substantial I/O time.

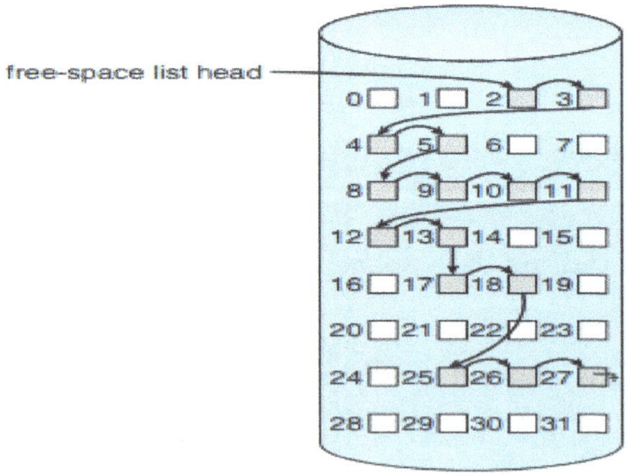

Figure 4.22 Linked free-space list on disk

4.10.3 Grouping

A modification of the free-list approach is to store the addresses of n free blocks in the first free block.

The first n-1 of these are actually free. The last one is the disk address of another block containing addresses of another n free block.

The importance of this implementation is that addresses of a large number of free blocks can be found quickly.

Counting

Another approach is several contiguous blocks may be allocated or freed simultaneously, particularly when contiguous allocation is used.

Thus, rather than keeping a list of free disk addresses, the address of the first free block is kept and the number n of free contiguous blocks that follow the first block.

Each entry in the free-space list then consists of a disk address and a count. Although each entry requires more space than would a simple disk address, the overall list will be shorter, as long as the count is generally greater than 1.

Space Maps

Oracle's ZFS file system was designed to encompass huge numbers of files, directories, and even file systems. On these scales, metadata I/O can have a large performance impact.

In its management of free space, ZFS uses a combination of techniques to control the size of data structures and minimize the I/O needed to manage those structures.

4.11 I/ O SYSTEMS

Overview

- Management of I/O devices is a very important part of the OS - so important and so varied that entire I/O subsystems are devoted to its operation.
- I/O Subsystems must contend with two trends:

  - The gravitation towards standard interfaces for a wide range of devices, making it easier to add newly developed devices to existing systems, and the development of entirely new types of devices, for which the existing standard interfaces are not always easy to apply.
- **Device drivers** are modules that can be plugged into an OS to handle a particular device or category of similar devices.

4.11.1 I/O Hardware

- I/O devices can be roughly categorized as storage, communications, user-interface, and other
- Devices communicate with the computer via signals sent over wires or through the air.
- Devices connect with the computer via **ports**, e.g. a serial or parallel port.
- A common set of wires connecting multiple devices is termed a **bus-** include rigid protocols for the types of messages.

There are three of the four bus types commonly found in a modern PC:

1. The **PCI bus** connects high-speed high-bandwidth devices to the memory subsystem (and the CPU.)
2. The **expansion bus** connects slower low-bandwidth devices, which typically deliver data one character at a time (with buffering.)
3. The **SCSI bus** connects a number of SCSI devices to a common SCSI controller.
4. A **daisy-chain bus,** (not shown) is when a string of devices is connected to each other like beads on a chain, and only one of the devices is directly connected to the host.

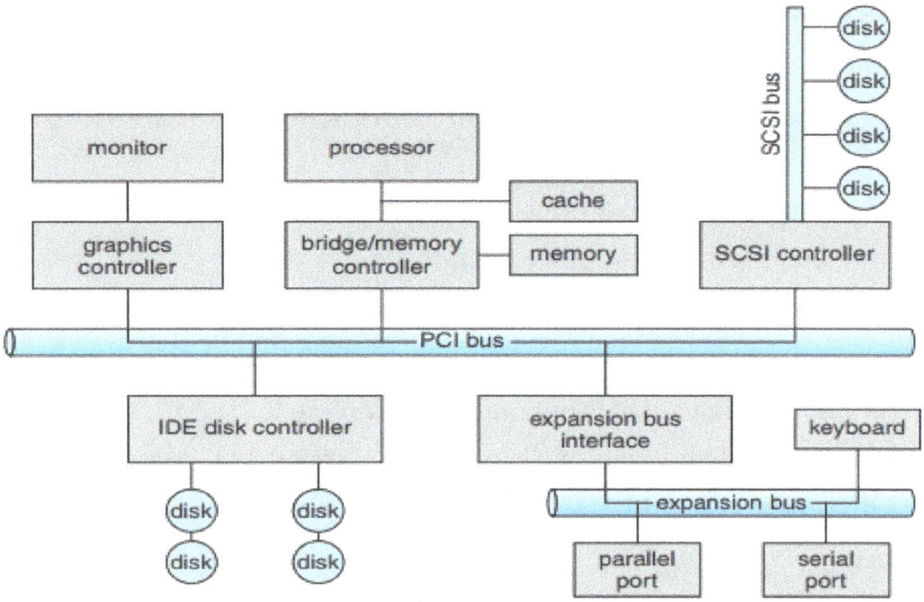

Figure 4.23 A typical PC Bus Structure

a) One way of communicating with devices is through **registers** associated with each port. Registers may be one to four bytes in size, and may typically include (a subset of) the following four:

1. The **data-in register** is read by the host to get input from the device.
2. The **data-out register** is written by the host to send output.
3. The **status register** has bits read by the host to ascertain the status of the device, such as idle, ready for input, busy, error, transaction complete, etc.
4. The **control register** has bits written by the host to issue commands or to change settings of the device such as parity checking, word length, or full-versus half-duplex operation.

I/O address range (hexadecimal)	device
000–00F	DMA controller
020–021	interrupt controller
040–043	timer
200–20F	game controller
2F8–2FF	serial port (secondary)
320–32F	hard-disk controller
378–37F	parallel port
3D0–3DF	graphics controller
3F0–3F7	diskette-drive controller
3F8–3FF	serial port (primary)

Figure 4.24 Device I/O port locations on PCs (partial)

b) Another technique for communicating with devices is **memory-mapped I/O.**

- In this case a certain portion of the processor's address space is mapped to the device, and communications occur by reading and writing directly to/from those memory areas.
- Memory-mapped I/O is suitable for devices which must move large quantities of data quickly, such as graphics cards.
- Memory-mapped I/O can be used either instead of or more often in combination with traditional registers. For example, graphics cards still use registers for control information such as setting the video mode.
- A potential problem exists with memory-mapped I/O, if a process is allowed to write directly to the address space used by a memory-mapped I/O device.

4.11.2 Polling

- One simple means of device **handshaking** involves polling:
  1. The host repeatedly checks the **busy bit** on the device until it becomes clear.
  2. The host writes a byte of data into the data-out register, and sets the **write bit** in the command register (in either order.)
  3. The host sets the **command ready bit** in the command register to notify the device of the pending command.
  4. When the device controller sees the command-ready bit set, it first sets the busy bit.
  5. Then the device controller reads the command register, sees the write bit set, reads the byte of data from the data-out register, and outputs the byte of data.
  6. The device controller then clears the **error bit** in the status register, the command-ready bit, and finally clears the busy bit, signaling the completion of the operation.
     - Polling can be very fast and efficient, if both the device and the controller are fast and if there is significant data to transfer.
     - It becomes inefficient if the host must wait a long time in the busy loop waiting for the device or if frequent checks need to be made for data that is infrequently there.

4.11.3 Interrupts

- Interrupts allow devices to notify the CPU when they have data to transfer or when an operation is complete, allowing the CPU to perform other duties when no I/O transfers need its immediate attention.
- The CPU has an **interrupt-request line** that is sensed after every instruction.
  - A device's controller **raises** an interrupt by asserting a signal on the interrupt request line.
  - The CPU then performs a state save, and transfers control to the **interrupt handler** routine at a fixed address in memory. (The CPU **catches** the interrupt and **dispatches** the interrupt handler).
  - The interrupt handler determines the cause of the interrupt, performs the necessary processing, performs a state restore, and executes a **return from interrupt** instruction to return control to the CPU. (The interrupt handler **clears** the interrupt by servicing the device).

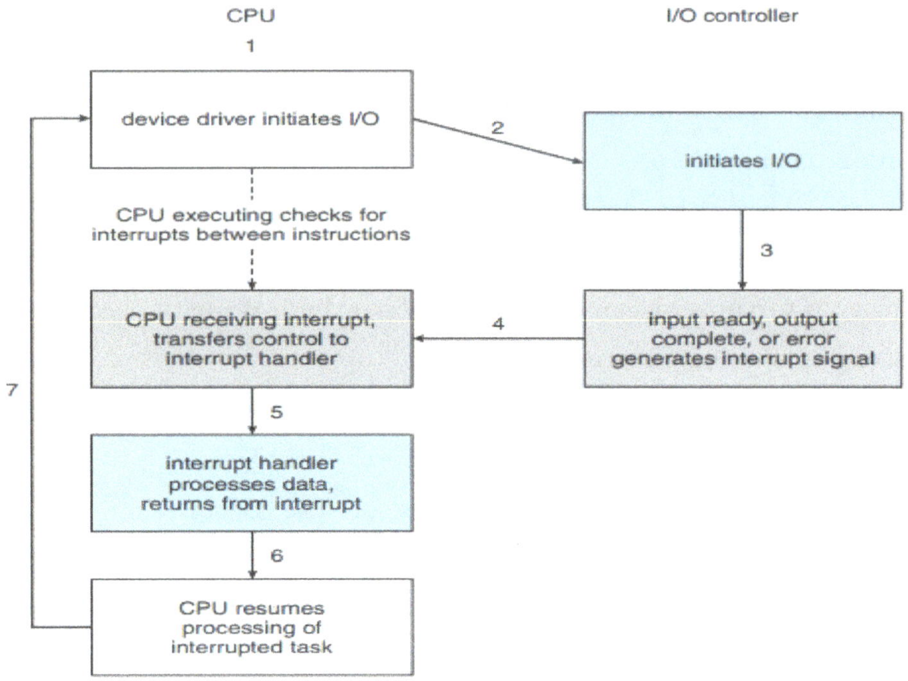

Figure 4.25 Interrupt-driven I/O cycle

- The above description is adequate for **simple interrupt-driven** I/O, but there are **three needs** in modern computing which complicate the picture:
    1. The need to defer interrupt handling during critical processing,
    2. The need to determine which interrupt handler to invoke, without having to poll all devices to see which one needs attention, and
    3. The need for multi-level interrupts, so the system can differentiate between high- and low-priority interrupts for proper response.
- These issues are handled in modern computer architectures with **interrupt-controller** hardware.
    - Most CPUs now have two interrupt-request lines:
        - **non-maskable** for critical error conditions and
        - **maskable,** that the CPU can temporarily ignore during critical processing.
    - The interrupt mechanism accepts an **address,** which is usually one of a small set of numbers for an offset into a table called the **interrupt vector.** This table holds the addresses of routines prepared to process specific interrupts.
    - The number of possible interrupt handlers still exceeds the range of defined interrupt numbers, so multiple handlers can be **interrupt chained**. Effectively the addresses held in the interrupt vectors are the head pointers for linked-lists of interrupt handlers.
    - Modern interrupt hardware also supports **interrupt priority levels**, allowing systems to mask off only lower-priority interrupts while servicing a high-priority interrupt, or conversely to allow a high-priority signal to interrupt the processing of a low-priority one.
- At boot time the system determines which devices are present, and loads the appropriate handler addresses into the interrupt table.
- During operation, devices signal errors or the completion of commands via interrupts.
- Exceptions, such as dividing by zero, invalid memory accesses, or attempts to access kernel mode instructions can be signaled via interrupts.
- Time slicing and context switches can also be implemented using the interrupt mechanism.

The scheduler sets a hardware timer before transferring control over to a user process.

When the timer raises the interrupt request line, the CPU performs a state-save, and transfers control over to the proper interrupt handler, which in turn runs the scheduler.

The scheduler does a state-restore of a **different** process before resetting the timer and issuing the return-from-interrupt instruction.

- A similar example involves **the paging system for virtual memory**
  - A page fault causes an interrupt, which in turn issues an I/O request and a context switch, moving the interrupted process into the wait queue and selecting a different process to run.
  - When the I/O request has completed, then the device interrupts, and the interrupt handler moves the process from the wait queue into the ready queue.
- System calls are implemented via **software interrupts or traps.**
  - When a program needs work performed in kernel mode, it sets command information and data addresses in certain registers, and then raises a software interrupt.
  - The system does a state save and then calls on the proper interrupt handler to process the request in kernel mode.
  - Software interrupts generally have low priority, as they are not as urgent as devices with limited buffering space.
- Interrupts are also used to **control kernel operations**, and to schedule activities for optimal performance. **For example**, the completion of a disk read operation involves **two** interrupts:
  - A high-priority interrupt acknowledges the device completion, and issues the next disk request so that the hardware does not sit idle.
  - A lower-priority interrupt transfers the data from the kernel memory space to the user space, and then transfers the process from the waiting queue to the ready queue.
- The Solaris OS uses a **multi-threaded kernel** and priority threads to assign different threads to different interrupt handlers.

- This allows for the "simultaneous" handling of multiple interrupts, and the assurance that high-priority interrupts will take precedence over low-priority ones and over user processes.

4.11.4 Direct Memory Access

- For devices that transfer large quantities of data (such as disk controllers), it is wasteful to tie up the CPU transferring data in and out of registers one byte at a time.
- Instead this work can be off-loaded to a special processor, known as the **Direct Memory Access, DMA, and Controller.**
- The **host issues** a command to the DMA controller, indicating the location where the data is located, the location where the data is to be transferred to, and the number of bytes of data to transfer.
- The **DMA controller handles the data transfer**, and then interrupts the CPU when the transfer is complete.
- Handshaking between DMA controllers and their devices is accomplished through two wires called the **DMA-request** and **DMA-acknowledge** wires.
- While the DMA transfer is going on the CPU **does not have access to the PCI** bus, but it does have access to its internal registers and primary and secondary caches.
- DMA can be done in terms of either physical addresses or virtual addresses that are mapped to physical addresses. The latter approach is known as **Direct Virtual Memory Access, DVMA,** and allows direct data transfer from one memory-mapped device to another without using the main memory chips.
- Direct DMA access by user processes can speed up operations, but is generally forbidden by modern systems for security and protection reasons. (i.e. DMA is a kernel-mode operation.)

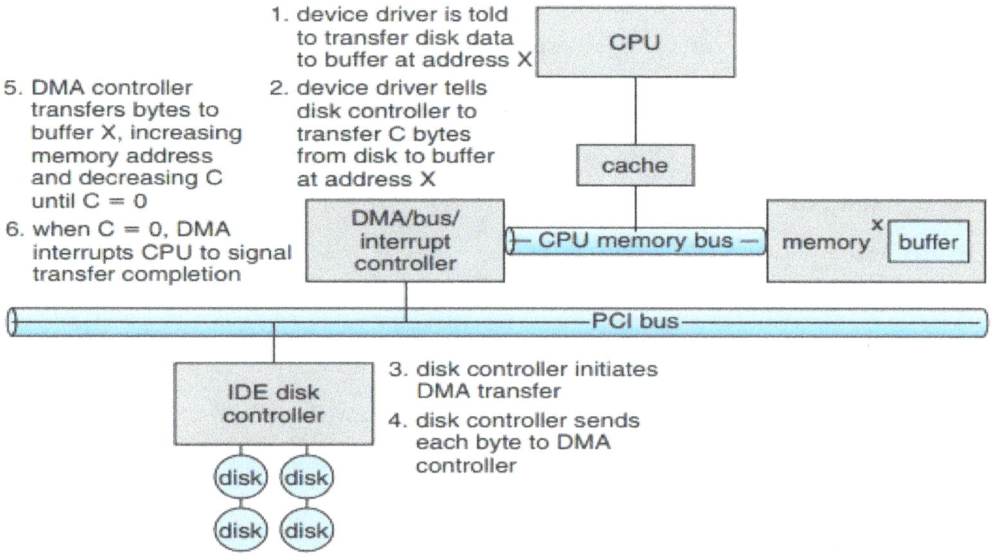

Figure 4.26 Steps in a DMA transfer

REVIEW QUESTIONS

PART-A

21. List the major components of magnetic disks.

Magnetic disks provide the bulk of secondary storage for modern computer systems. It consists of

- Disk platter
- Read-Write head
- Disk arm
- Tracks and Sectors
- Cylinder

22. Define transfer rate in magnetic disk.

The transfer rate is the rate at which data flow between the drive and the computer.

23. Define random access time or position time.

The random access time consists of two parts:

- **Seek time**- the time necessary to move the disk arm to the desired cylinder
- **Rotational Latency**-the time necessary for the desired sector to rotate the disk head.

24. What is meant by Input/Output bus?

A disk drive is attached to a computer by a set of wires called an I/O bus. Several kind of buses are available:

- Advanced Technology Attachment (ATA)
- Serial ATA(SATA), etc,.

25. What is meant by controller?

The data transfers on a bus are carried out by special electronic processors called controllers. There are two types of controllers:

- **Host Controller**-the controller at the computer end of the bus.
- **Disk Controller**-built into each drive.

26. List the methods to handle the bad blocks in disk management.

The methods to handle the bad blocks (a defective block) is:

- Sector sparing (or) forwarding
- Sector slipping

27. Define file.

A file is a named collection of related information that is recorded on secondary storage. The information in a file is defined by its creators. The types of information were:

- Source or executable programs
- Numeric or text data
- Photos
- Music, etc,.

28. List the attributes of a file.

A file attributes vary from one operating system to another but it commonly consists of the following:

- **Name**-the symbolic file name is the only information in human readable form
- **Identifier**- A unique tag, usually a number, identifies the file within the file system; it is the non-human readable form.
- **Type**-information needed for systems that support different types of files
- **Location**-This information is a pointer to a device and to the location of the file on that device.
- **Size**-current size of the file
- **Protection**-access-control information
- **Time, date and user identification**

29. What are the various file operations available?

A file is an abstract data type which provides the following operations:

- Creating a file-creates a new file
- Writing a file- write into the file
- Reading a file- read the content of a file
- Repositioning within a file (or) file seek-pointer is repositioned to the given value

- Deleting a file
- Appending a file
- Renaming a file
- Open and Close a file

30. List the various operations to be performed on a directory.

The directory can be viewed as a symbol table that translates file names into their directory entries. The various operations to be performed on directory includes:

- Search for a file
- Create a file
- Delete a file
- List a directory
- Rename a file
- Traverse the file system

31. What is meant by absolute and relative path name?

Absolute path name begins at the root and follows a path down to the specified file. Relative path name defines a path from the current directory.

32. List the technologies used in the evolution of network and file technology.

The technologies used in the evolution of network and file technology are:

- First method-Manual transfer
- Second method-Distributed file system
- Third method-World Wide Web
- Fourth method-Cloud Computing

33. What is meant by anonymous access?

Anonymous access allows a user to transfer files without having an account on the remote system.

34. List the protection mechanisms available for limiting the types of file accessing.

Protection mechanisms provide controlled access by limiting the types of file access that can be made. Several different types of operations may be controlled:

- Read
- Write
- Execute
- Append
- Delete
- List

35. Draw a neat diagram of the structure of a file.

Application programs

⇩

Logical file system

⇩

File organization module

⇩

Basic file system

⇩

I/O control

⇩

devices

36. List the structure available in On-Disk file system implementation.

The On-disk file system implementation includes the structures such as:

- Boot control block
- Volume control block
- Directory structure
- A per-file file control block

37. List the structure available in In-Memory file system implementation.

The In-memory file system implementation includes the structures such as:

- Mount table
- Director structure cache
- System-wide open-file table
- Per-process open-file table
- Buffers

38. What are the methods available for the directory implementation?

The methods available for the implementation of the directory are:

- Linear list
- Hash table

PART-B

1. Write short notes on the magnetic disks.

   Explain the following disk scheduling algorithms with suitable example:

   a) First-Come First-Server (FCFS) scheduling

   b) Shortest-Seek-Time-First (SSTF) scheduling

   c) SCAN scheduling

   d) Circular SCAN (C-SCAN) scheduling

   e) LOOK scheduling

2. Write short notes on disk management.

   Explain in detail about the various schemes for defining the logical structure of a directory.

   Write short notes on

   a) File sharing

   b) File protection

   c) File system structure

4. Explain in detail about the file system implementation.
5. Explain in detail about the various allocation methods available for disk allocation.
6. Explain about the free-space management in detail.
7. Explain in detail about the Input / Output systems.

UNIT-5

CASE STUDY

5.1 Linux System - Basic Concepts:

- Linux is a Modern, face operating system based on UNIX standards.
- Developed in the year **1991** by **Linux Torvalds,** as a small kernel with the major design goals of UNIX compatibility.
- It has been designed to run efficiently and reliably on common PC hardware, but also runs on a variety of other platforms.
- Linux Wish Functions includes
  - Kernel
  - Applications
  - Management tools.

Design Principles:

Linux is a Multiuser, Multitasking system with a full set of UNIX compatible tools.

- Main Design goals are Speed, efficiency and Standardization.

Components of a Linux System:

<table>
<tr><th>System Management Programs</th><th>User Processes</th><th>User Utility Programs</th><th>Compilers</th></tr>
<tr><td colspan="4">System Shared Libraries</td></tr>
<tr><td colspan="4">Linux Kernel</td></tr>
<tr><td colspan="4">Loadable Kernel Modules</td></tr>
</table>

Figure 5.1 Components of a Linux System

- The Kernel is responsible for maintaining the important abstractions of the OS.
  - Kernel Code executes in Kernel mode with full access to all the physical resources of the computer.
  - All kernel code and data Structures are kept in the same single address space.
- The **System Libraries** define a Standard set of functions through which applications interact with the kernel.
- The system utilities perform individual specialized management tasks.

Kernel Modules

Modules are pieces of code that can be loaded and unloaded into the kernel upon demand. They extend the functionality of the kernel without the need to reboot the system.

For example, one type of module is the device driver, which allows the kernel to access hardware connected to the system. Without modules, we would have to build monolithic kernels and add new functionality directly into the kernel image. Besides having larger kernels, this has the disadvantage of requiring us to rebuild and reboot the kernel every time we want new functionality

A **Kernel Module** implement a device driver, a file system or a networking protocol. It allows a Linux System to be set up with a Standard, minimal Kernel, without any extra device drivers built in.

Basic Linux Commands:

Linux is an Operating System. It is similar to the UNIX OS.

Linux Commands:

1. LS => This command is used to list all the files in the current working directory.

   Eg: $ls
2. Rm => This command is used to remove the file from the directory.

   Eg: $rm filename
3. Rmdir => This command is used to remove directory. But it should not contain any file.

   Eg: $rmdir directory name

4. Cat => This command is used to display the content of file.

   Eg: $cat >filename(i.e)used to write the content into the file.

5. CD => This command is used to change the current working directory.

   Eg: $cd

6. Mv => This command is used to move the content from the Source file to destination file. It is also called rename command.

   Eg: $mv sourcefile destinationfile

7. Who => This command is used to display the users who are currently logged on.

   Eg: $who

8. Who am I => This command is used to display the information about the users.

   Eg: $who am i

9. Mkdir => This command is used to create a new directory.

   Eg:$mkdir directoryname

10. Cp => This command is used to copy the content of Source file to the destination file.

    Eg: $cp source_file destination_file

11. Grep => This command is used to search a word or character or sentence in a file.

    Eg: $grep word filename

12. pwd => This command is used to display current working directory.

    Eg: pwd

13. **chmod** => This command is used to change the file access permission.

14. **Clear** => This command is used to clear the terminals.

15. **W** => This command is used to print the System users.

Linux Directory Structure:

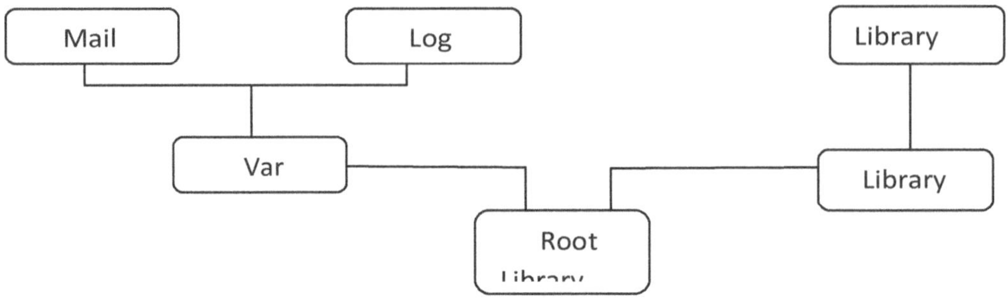

Figure 5.2 Linux Directory structure

5.2 System Administration:

- System administration includes common issues and tips for Linux system adminstration.

5.2.1 Role of System Administration:

The System Administrator has the ability to

- Work Independently.
- Think analytically and a problem solver.
- Communicate effectively, both in waiting and in speaking , with customers,co-workers and various business contacts in a professional manner.
- Work completely and accurately under time constraints and deadlines.
- Obtain and maintain a valid driver's license.
- Read,analyze and interpret reports.
- Provide excellent customer service.
- Work in a safe and effective manner.

The common issues to be concerned by the system administrator are

- Basic Configuration
- Set the Host Name
- Set the Time Zone
- Network Diagnostics
- Using the ping command.
- Using the Trace route command.

- Using the mtr command.
- System diagnosis
- Check current memory usage.
- Monitor I/O usage with VMstat
- Monitor Processes, memory and CPU usage with HTOP.

1.Basic Configuration:

It includes basic steps and issues encountered during the begin of system configuration.

2. Set the HostName:

The following instructions are used to set HostName

- HostName=> shows short HostName
- HostName_f => shows fully Qualified Domain Name(FQDN).

3. Set the Time Zone:

When Setting the time zone of server, the best way is to set the time zone of bulk of users.

- If unsure about the best Time Zone universal coordinated time (or) UTC (Greenwich Mean Time) Used.
- By default, Linode base installs are set to **Easten Standard Time**.
- To change the timezone manually, find a proper zone file in**/usr/share/zone info/and link that file to /etc/local time.**
- To change the timezone in Webian and ubuntu Systems.

  dpkg-reconfigure tzdata.
- In arch Linux

/etc/rc.conf =>**Provides a list of IP address with host name.**

TimeZone="America/New-york".

4. Network Diagnostics

Ping command, trace route command and mtr command are used to diagonose network problems.

- **Using the ping command:**
  - Ping command test the connection between the local m/c and a remote address on m/c.
  - The toll command "pings" google.command and 74.125.67.100 ping google.com ping 74.125.67.100

These commands send a bit of data (i.e 1 CMP pkt) to the remote host and wait for a response.

If the system is able to make a connection,for every packet it will report at the "Round trip Time".

Important Points:

1. **Packet loss** or the discrepany between the number of packets send and the number of that return successfully.
2. **Round Trip Time** ping is useful as an a informal diagnostic tool to measure **pt-to-pt network latency** , and as a network connection tool.

- **Using the trace route command:**

  TraceRoute provides a respect on the path that the packets take to get from the locals m\c to the remote m\c.

Route information is useful , for troubleshooting a network issue.

i.e

=>If packet loss is first few hops->problem is related to user's LAN (&) Internet service provider(ISP).

=>If the packet loss near the end of route->problem related to the Server's connection

Ex: Traceroute to google.com(74.125.53.100), 30hops max, 40bytes packet.

- **Using the mtrcommand:**

  Mtr provides information about the route that internet traffic takes between the local system and a remote host.

Also provides information about the **Round Trip Time.**

Ex:host: squire.example.com Loss% snt Last average best west st.dev

5. System Diagnostics:

=>**In Linode any issue neither related to networking nor** another easily diagnosable application issue,the toll tools used for the better diagnose.

=>If the problem is with memory usage ,

- **Check Current Memory Usage:**

=>To see the memory usage of the system at the current node, the following command is used.

Free –m

=>This command will generate the resembles as follows:

Total used free shared buffer cached memory: 1002 956 46 0 171 357 -/+ buffers/

Cache:427 575

Swap:127 39 88.

- **Monitor IO Usage with Vmstat:**

=>The Vmstat tool provides information about memory swap utilization, I/O wait and System activity.

=>It is particularly useful for diagnosy I/O Related issue.

=> The following command is issued if there is an I/O issue: **Vmstat 1 20-runs a Vmstat every second,20 times.**

- **Monitor Processor,Memory and CPU usage with Htop**

=>If there is a need of more organized and real time view of the current state of the system, a htop tool is used.

=>To install htop , the following commands are issued .

1. apt-get install htop
2. yiem install htop
3. pacman –s htop
4. emerge system- Process/htop.

=>At the command line, issue the following command

htop

=>can be quit by anytime by pressing the FIO or 4 keys

5.2.2 Requirements of a System Administrator

The System administrator responsible for the following:

1. User administration-Setup and maintaining a account.
2. Verify the peripherals.
3. Monitor System performance and network communication.
4. Create File Systems, backup and recover policy.
5. Install Software.
6. Update the new version of OS and application software.

5.3 Setting Up A Linux Multi Function Server:

Following are the instruction provided a good basics for getting a grip on Linux.

- **Server requirements:**
  - ⇨ Distribution of linux to configure a web server: Debian.
  - ⇨ To setup linux internet server:Need internet connection and a static IP address.
- **System requirements:**
  - ⇨ Pentium III CPU.
  - ⇨ Minimum of 256MB of RAM.
  - ⇨ 10GB harddrive.
  - ⇨ Newer CPU and additional memory for better performance.
  - ⇨ Debian website:http://www. Debian.org-provides dowmloadable CD images.
- **Installing Debian:**

  Steps to Proceed for installing debian s/w are:

1. Boot into the Debian CD-ROM disk to get the login screen.
2. Installable guide will provide a service of Installation screens.
3. Screen called "configure the N/W" Will be displayed to suggests the n/w with DHCP.
4. Ask for the details such as hostname of the server, a domain name , a gateway , an IP address, a n/w and a name server.
5. A/F registering the domain and a static IP address , Ready to go.
6. A number of registrars will be listed.
7. To obtain a domain name initially, two registered DNS servers are needed.

⇨ A primary DNS server.

⇨ A backup on secondary DNS server.

8. A/F doing all the above configuration, continue with the installation tasks to complete the base system.

- **Logging in Remotely**
  - A/f the installation was successful, log into the server from a remote console on the desktop
  - The further administration were can be carried out from another system because a secure server should maintained.
  - On the remote m/c, need only can SSH client,which virtually all linux distributions can be downloaded for other operating systems.

- **Configuring the Network**

There are 3 options for configurations.

- If DHCP used for Debian Installations then configure the server with static IP addresses, then perform testing.
- If public IP addresses used, and configured as static, then can skip to next sections.
- If Debian installed with a DHCP client from router (or) Internet Service provider, then need to reconfigure networking.
  - To change the settings, static IP addresses used , and need to become root and then edit the file
  - Example: IP address 70.153.258.42
  - A/f editing the etc/network /interfaces file,restart the network by entering #/etc/init.d/networking restart.
  - Then, need to edit /etc/resolve.conf and add nameservers to resolve internet hostnames to their corresponding IP addresses.
  - Then setup a minimal DNS server.
  - The other nameservers, should specify the IP addresses of the DNS servers offered by our ISP.

i.e., our resolve configuration looks as follows

search server

name server:70.153.258.42

name server:70.253.158.45

name server:151.164.1.8

- use only DNS servers that work with our domain site
- then edit /etc/hosts and add our IP address.

127.0.0.1 localhost.localdomain localhost server

70.153.258.42 sever1.centralsoft.orgserver1

- Then set the hostname, enter the following commands

#echo server1.Centralsoft.org>/etc/hostname

#/bin/hostname – F/etc/hostname

#/bin/hostname-F/etc/hostname

- Substitute your domain name for server1.centralsoft.org
- Then,verify the congigured host name by running the host name command
- $host name-server1
- $host name-server1.centralsoft.org
- Then, if result comes, ok, otherwise look in the /etc/hostname file

- **Changing the Default Debian Packages**

  With the packages of debian during the installation, need to make some changes in order to use postfix

  - The Debian team has chosen to install, by default services appropriate for a LAN
  - But for the server to be placed on the Internet,need some changes to be ,ade with the existing.

5.4 DOMAIN NAME SERVER (DNS)

Domain Name Server (DNS) is the service used to convert human readable names of hosts to IP addresses

Three main components of DNS

- Resolver
- Name Server
- Database of resource records (RRs)

DNS Server

If any computer registered to join the DNS(Domain name system). A DNS server runs special purpose networking software, features a public IP address and contains a database of n/w names and addresses for other Internet hosts.

Resolver

- The DNS is a large database which resides on various computers and it contains the names and IP addresses of various hosts as the internet and various domains
- The Domain Name System is used to provide information to the Domain Name Service to used when queries are made.
- A Resolver is a program or part of operating system that resides an user workstation, used to access to the domain name databases.
- The resolver will send requests to the name servers to return information requested by the user
- The requesting computer tries to connect to the name server using its IP address

Name servers

Three types of name server, namely:

1) Primary master – builds its database from files that were preconfigured on its hosts
2) Secondary master – provide information to resolver by getting the information from the primary.
3) Caching name server – gets all its answers to queries from other name servers and serves the answers. It is a non-authorative server.

Database of Resource Records:

A database is made up of records and the DNS. Common resource record types in the DNS database are

- Hosts IP address
- PTR-hosts domain name
- CNAME-hosts canomical name
- MX- hosts or domain's mail exchanges
- NS- host's or domain's name server

- SOA-Indicates authority for the domain
- TXT-generic text record
- SRV-service location record
- RP-responsible
- HINFO-Host information record withCPU type and OS

When a resolver requests information from file server, the DNS query message indicates one of the preceding types

5.5 SETTING UP A LOCAL NETWORK SERVICES

Local Network Services includes

- Distributed file system
- Introduction to Samba
- Configuring the network
- Installing DHCP
- Starting your DHCP service
- Providing static IP address

5.5.1 Distributed File systems

In distributed file systems, user open, read and write files stored in computers other than their own.(i.e.,) file and directories can be sharing and users can access in the remote machines called shares

5.5.2 Introduction to Samba

- SAMBA is the most popular free-software prospects to implement Microsoft file sharing in non-microsoft systems.
- It has significant support for Windows and Linux desktops and an MAC OS X.
- It also play a file sharing role in simpler environments like small offices and larger organizations.
- With the use of Samba , desktop users can share their printers and files with others, without those others having to authenticate.
- Stranger, machine level security policies can be implemented to shield that machine from other users without compromising its ability to access the resources of the peer-to-peer network.

5.5.3 Configuring the Network

- The Linux system looks the same as a windows system.
- In home window, left hand side shows all folders, under My Linux. Right hand side shows the shared documents folders.
- For network setup standard wiring, Ethernet connections and installation of the firewall and modem are done.
- Standard setup procedures are used on Windows 98 machines.
- The systems uses DHCP to obtain IP addresses, DNS servers, and routes to a gateway.
- The router provided DHCP services and a private Internet address scheme using a class C network(192.168.0.0 through 192.168.0.255).
- Other Linux distributions, such as Fedora and Ubuntu, also offer easy tools for setting up Windows8 file sharing.

5.5.4 Installing DHCP

- To get started with DHCP, DHCP server need to be installed.
- The current version of the package is dhcp_3.0.3_28.i386
- Debian users can install dhcp3 server package and edit configuration file.

/etc/dhcp3/dhcpd.conf

- Once DHCP installed, configure DHCP **in /etc/dhcpd.conf.**
- As a first step, copy the **file /usr/share/doc/dhcp/dhcpd.conf.sample** to **/etc/dhcpd.conf.next.**
- Edit the file to fit in the network.

Syntax:

- Pool 3 IP address available for the users in the subset of LAN
- For reserved private class C network 192.168.1.0 which provides 254 nodes. This netmask must match with one LAN.
- Gateway address, optim routers,caching name server, optim domain name servers are specified. A single server with 2 networks and acts as a gateway to LAN. One card represented as eth0, has an address in Internet, while the other card eth1 has an address in private network.
- Enabled BIND in caching mode to function as the networks DNS server

Starting DHCP Service

- Some DHCP services require a dhcpd leases file
- **touch command** is used to create an empty file in the same directory **dhcpd.conf** file:

  #touch/var/lib/dhcp/dhcpd.leases

- **To Start DHCP Server**
  - To check whether the configuration is correct

[root@host2n]#service dhcpd start

Starting dhcpd:[ok]

[root@host2n]#

- **To Test Whether DHCP Process Is Running**

#ps aux/grep dhcpd

Root 9028 0.00.0 2552 636 ss

09:40 0:00/usr/sbin/dhcpd

Check configuration command used to get DHCP to start at boot time

5.5.5 Providing Static IP Addresses

Workstations run fine with dynamic addresses but servers usually benefit from static addresses

Steps:

1. Set up the sabnet, broadcast address and routers:

   Subnet 192.168.1.0 netmask 255.255.255.0

   Optim broadcast add 192.168.1.255

   Optim routers 192.168.1.1

2. Add a host section for each machine
   - For this hardware address i.e., MAC address is needed for each network card.
   - **IP configuration** command is used.

Example:

#ethernet MAC address as follows

Host laser_printer{

Hardware Ethernet 08:00:2b:4c:59:23;

Fixed address 192.168.1.10;

```
}
Host1.centralsoft.com
{
Hardware Ethernet 01:0:c0:2d:8c:33;
Fixed_address 192.168.1.5;
}
```

5.6 VIRTUALIZATION

5.6.1 Basic Concepts

- Virtualization is a concept that has gained popularity,because of the successful company Vmware (http://www.vmware.com)
- And the open source project xen http://www.c1.cam.ac.uk/research/srg/netos/new)
- It refers to one piece of hardware running multiple kernels in top of a lower layer of software that manages their access to the hardware.
- Each kernel, called a gnest, acts as if it has the whole processor to itself
- The different guests are isolated from each other much more than processes are isolated within a single operating system.
- This isolation provides security and robustness, because a failure or compromise in one guest doesn't affect others.
- The virtualization layer performs many functions of an OS, managing access to processor time, devices, and memory for each guest.

- **Virtual Machine (VM)**

A software implementation of a machine that executes programs like a physical machine Types of virtual machine categories:

1. **System Virtual Machine**-Hardware Virtual Machine Provides a complete system platform environment which supports the execution of a complete operating system(OS)
2. **Process Virtual Machine**-Application Virtual Machine, Provides a platform independent primary environment that abstracts away of the underlying hardware or OS from software or application runtime.

Example:

- Hardware virtual machine: VMWare, XEN, Virtual Box.
- Application virtual machine:java Virtual M/c, .net framework

Test before we go:

- We can have virtual machine just like an experiment box. Able to test applications in different trends of system in same machine.
- Its east to destroy, rebuild or backup whole testing environments-Virtual machine.

Server Consolidation

- Many small physical servers could be replaced by virtual machine, runs in virtualization environment provided by one powerful physical server
- Each OS running in a physical server becomes converted to a distinct OS running inside a virtual machine.
- The large server can "host" many such "guest" virtual machines.

Increase hardware Utilization:

- To increase the utilization of costly hardware resources such as cpu, memory or even like storage space.
- Rapid provisioning , dynamic fault tolerance.

Example:

- Computing task , can be utilized down to core of cpu level.
- Big system memory can partition as smaller portions for legacy system runs within a virtual machine.

Traditional Vs virtual Architecture:

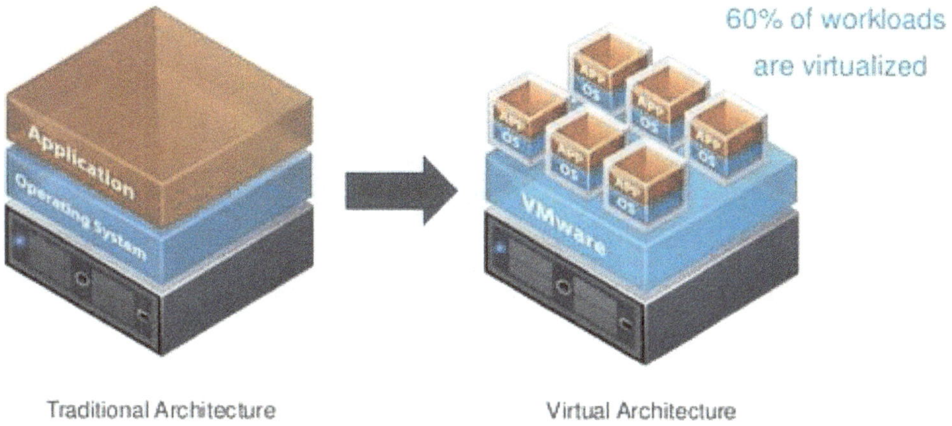

Figure 5.3 Comparisons of traditional and Virtual Architecture

5.6.2 Setting Up Xen:

Xen Architecture:

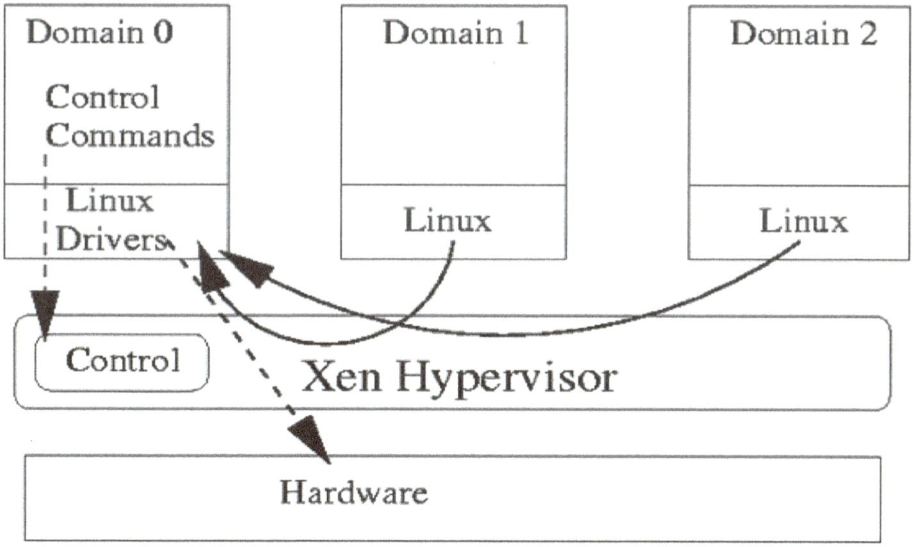

Figure 5.4 Xen Architecture

Layers of Xen Architecture:

- Virtual machine layer.
- Hypervisor layer
- Hardware / physical layer.

Hypervisor layer:

Thin layer of software that runs an top of the hardware. The xen hypervisor gives each virtual machine a dedicated view of the hardware.

Virtual machine layer:

Operating System hosted on the hypervisor and appearing to the user as a separate physical computer. However , the machine shares physical resources with other virtual machines, and it is portable because the virtual machine is abstracted from the physical hardware.

XEN-1:

Xen is a virtual machine for IA-32 (X 36, X86-64), IA-64 and power pc 970 architecture. It allows several guest operating systems to be executed on yhe same computer hardware concurrently.

Xen was initially created by the university of Cambridge, computer laboratory and is now developed and maintained by the xen community as free software as well as citrix Xen server commercial version variant.

A Xen System is structured with the xen hypervisor as the lowest and most privileged layer. Above this layer are "one " or "more" guest os,which the hypervisor schedule across the physical CPUs.

The first guest OS called Xen terminology. a.k.a "Domain 0" (dom 0) , is boosted automatically when the hypervisor boots and given special management privileges and direct to the physical hardware.

System Administrator can log into dom 0 in order to manage any further guest operating system , called "domain U" (dom U) in Xen technology.

Xen-2:

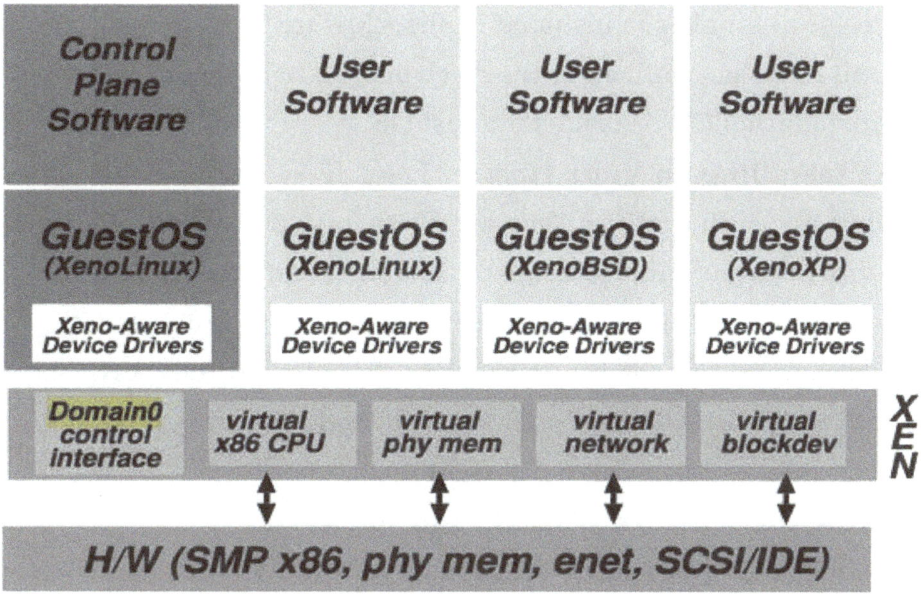

Figure 5.5 Architecture of Xen- 2

5.7 VMWARE ON LINUX HOST AND ADDING GUEST OS:

Getting started with VMware workstation is simple. The key steps are:

1. Install the VMware workstation software as described below.
2. Start VMware workstation and enter (your) serial number.
3. Create a virtual machine using the new virtual machine wizard.
4. Install a guest OS with new virtual machine. (Using CD-ROM a floppy disks).
5. Install the VMware tools package in virtual machine for enchanced performance.
6. Start using virtual machine.

- **Before begin the following steps to be assured as available,**

1. A computer and host OS that meet the system requirtements for running VMware Workstation.
2. The VMware workstation installation software
   - ⇨ In the packaged distribution- software is in CD.
   - ⇨ In Electronic distribution- software is in the file,which is downloaded.

3. VMware Workstation serial number:
   - ⇨ The serial number is included in the VMware Workstation package or in the email message confirming the electronic distribution order.
4. The installation CD or disks for guest OS.

- **Before installing on linux Host:**

Before install and run Vmware Workstation check the following notes and make any necessary adjustments to thr configuration of host Os.

1. The real time clock function must be compiled into Linux kernel.
2. VMware workstation for Linux requires that the parallel portpc-style hardware option (CONFIH-PARPORT-PC) be built and loaded as a kernel module (ie it must be set to when the kernel is compiled).

- **Installing the VMware Workstation software:**

1. Log on to linux host with the username.
2. In a terminal window , become root and can perform initial installation steps.
3. Su-
4. Mount the VMware Workstation CD-ROM .
5. Change to the Linux directory on the CD.
6. Do one of the following.\

To use the RPM installer, run RPM specify the installation File.

rpm-uhv vmware-< xxxx >.rpm

Vmware - <xxxx>.rpm

Is the Installation file in the CD; in place of <xxxx> the file name contains numbers that correspond to the version and build.

- **Upgrading from VMware Workstation 3.0:**

Before install the RPM package need to remove the prebuilt modules RPM package included in the 3.0 release. To remove the modules, type the foll at a command prompt:

rpm –e VMware Workstation kernel Modules.

To use tar installer:

Copy a tar archive to harddisk and install.

(or)

May skip the steps for copyimg and unpacking the archive and install directly from the VMware- distub directory an the CD.

- Copy the tar archive to a directory an harddrive .

  cp vmware -<xxx>.tar.gz/tmp
- Change to the directory to which the file is copied. **cd/tmp**.
- Unpack the archive.

  tar zxf VMware -< xxx >.tar.gz
- Change to the installation directory.

  cd vmware –distrib
- Reen the installation script

  ./vmware – install.p1
- Accept the default directories for the binary files, library files , manual files , documentation files and init script.
- Run the configuration script.

  vmware – config.p1

- **If RPM installer used**

  Need to run this script separately from the command line.If installed using tar archive the installer offers to launch the configuration script.

- Use the script to reconfigure vmware workstation whenever kernel upgraded. If is not necessary to reinstall vmware workstation after kernel upgradation.
- Vmware –config.p1 to reconfigure the networking options for Vmware Workstation

  Ex : to add or remove host only networks.
- Press enter to read the end user license agreement (EULA).

REVIEW QUESTIONS

PART-A

39. What is meant by Linux? List its purpose.

Linux is a modern, free operating system based on UNIX standards. It was first developed as a small but self-contained kernel in 1991 by Linus Torvalder, with the major design goal of UNIX compatibility. It has been designed to run efficiently and reliably on common PC hardware, but also runs on a variety of other platforms.

40. What are the components of linux system?

The components of linux system are

System management programs	User programs	User utility Programs	Compilers
System shared libraries			
Linux Kernel			
Loadable Kernel modules			

41. List the basic shell commands in Linux.

Linux is an operating system which is similar to UNIX OS. The basic linux commands are:

Commands	Description	Commands	Description
ls	Command used to list all the files in the current working directory	cp	Command used to copy the content of source file to the destination file
rm	Command used to remove the file from the directory	cat	Command used to copy the content file
rmdir	Command used to remove directory	grep	Command used to search a word or character or sentence in a file
mkdir	Command used to create a new directory	pwd	Command used to display current working directory
CD	Command used to change the current working directory	chmod	Command used to change the file access permission
mv	Command used to move the content from the source to the destination file	clear	Command used to clear the terminals
who	Command used to display the users who are currently logged on	w	Command used to print the current system users
who am i	Command used to display the information about the user		

42. How will be the directory structure in Linux.

The directory structure in Linux is as follows:

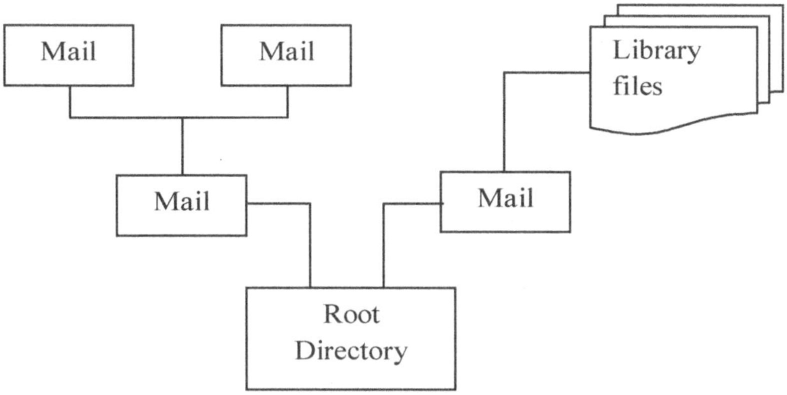

43. List the roles of a system administrator.

The system administrator has the ability to

- Work independently
- Think analytically and be a problem solver
- Work completely and accurately under time constraints and deadlines
- Provide excellent customer service
- Work in a safe and effective manner

44. What are the responsibilities of a system administrator?

The system administrator is responsible for the following:

- User administration to setup and maintaining account
- Verify the peripherals
- Monitor system performance and network communication
- Create file systems, backup and recover policy
- Install software
- Update the new versions of OS and application

45. What are the system requirements for setting up a Linux multifunction server?

The system requirements for setting up a Linux multifunction server are:

- Pentium III CPU
- Minimum of 256 MB of RAM
- 10 GB hard drive
- Newer CPU and additional memory-for better performance

46. List the components of domain name server.

The three main components of Domain Name Server are:

- Resolver
- Name server
- Database of resource records (RRs)

47. What are the types of name server available in DNS?

Three types of name server in DNS are:

- Primary master
- Secondary master
- Caching name server

48. What are the benefits of virtualization?

Virtualization has the following benefits:

- It replaces wasteful arrays of systems with better utilize systems.
- It simplifies administration because separate kernels with one application running on each are more secure and manageable than one kernel running many applications.
- It also maintains the environment in which documents were created, to meet regulatory requirements.
- Reduced hardware and complexity allows reduced staff.
- Virtualization may help reverse the trend of server sprawl.

PART-B

1. Explain the purpose, components and basic shell commands of Linux with suitable examples.
2. Explain the requirements of system administrator and the basic configuration in Linux in detail.
3. Explain in detail about setting up a Linux multifunction server.
4. Discuss in detail about the Domain Name System.
5. Explain the steps and procedure required for setting up a local network services in detail.
6. Write short notes on Virtualization.
7. Explain the steps and procedure required for setting up a XEN in detail.
8. Explain in detail about the VMware on Linux host and adding a guest OS.